When the Caribou
Do Not Come

When the Caribou Do Not Come

Indigenous Knowledge and Adaptive Management in the Western Arctic

EDITED BY BRENDA L. PARLEE

AND KEN J. CAINE

UBC Press • Vancouver • Toronto

26 25 24 23 22 21 20 19 18 17   5 4 3 2 1

Printed in Canada on FSC-certified ancient-forest-free paper (100% post-consumer recycled) that is processed chlorine- and acid-free.

Library and Archives Canada Cataloguing in Publication

When the caribou do not come : indigenous knowledge and adaptive management in the western Arctic / edited by Brenda Parlee and Ken Caine.

Includes bibliographical references and index.
Issued in print and electronic formats.
ISBN 978-0-7748-3118-5 (hardcover).–ISBN 978-0-7748-3119-2 (pbk.).–
ISBN 978-0-7748-3120-8 (PDF).–ISBN 978-0-7748-3121-5 (EPUB).–
ISBN 978-0-7748-3122-2 (Kindle)

  1. Caribou–Canada, Northern. 2. Wildlife management–Canada, Northern.
3. Traditional ecological knowledge–Canada, Northern. 4. Inuit–Hunting–Canada, Northern. 5. Indians of North America–Hunting–Canada, Northern. I. Parlee, Brenda, editor II. Caine, Ken, editor

QL737.U55W54 2017   599.65'80971   C2017-905708-1
                                                  C2017-905709-X

Canadä

UBC Press gratefully acknowledges the financial support for our publishing program of the Government of Canada (through the Canada Book Fund), the Canada Council for the Arts, and the British Columbia Arts Council.

This book has been published with the help of a grant from the Canadian Federation for the Humanities and Social Sciences, through the Awards to Scholarly Publications Program, using funds provided by the Social Sciences and Humanities Research Council of Canada.

Printed and bound in Canada by Friesens
Set in Garamond by Apex CoVantage, LLC
Copy editor: Robert Lewis
Proofreader: Caitlin Gordon-Walker
Indexer: Judy Dunlop
Cover designer: Martyn Schmoll
Cartographer: Eric Leinberger

UBC Press
The University of British Columbia
2029 West Mall
Vancouver, BC V6T 1Z2
www.ubcpress.ca

In memory of Teetł'it Gwich'in elder Elizabeth Colin:
"Women have voice for the caribou too"
With love to Scott, Eric, and Alex
for their support and patience on this journey
~ Brenda Parlee
and to the Délı̨nę Got'ı̨nę,
who have taught the deeper meaning of caribou
~ Ken Caine

Contents

Foreword

Fikret Berkes

In his classic book *The Wind and the Caribou* (1953), Erik Munsterhjelm quotes an ancient Dene saying, "No one knows the way of the winds and the caribou" (p. 97). Barren-ground caribou, one of the most important wildlife species of the North American Subarctic and Arctic, travel in very large numbers when they come, but they are unpredictable. Ancient hunters must have perceived caribou as undepletable but, alas, also as impossible to foretell. The failure of caribou migrations and the starvation of the Inuit people that followed have been embedded in Canadian culture through Farley Mowat's *People of the Deer* (1952). Even though the veracity of Mowat's particular story has been questioned, this does not make the hardship caused by the disappearance of caribou any less real.

Starvation is no longer common, but the phenomenon of disappearing caribou is still with us. There was one caribou crisis in the 1970s, and we seem to be in the middle of another one at the time of the publication of this book. Is the crisis real? Has there really been a sharp decline of the various populations? How can the observed changes be explained? As this book shows, these questions are almost as perplexing to current-day hunters, managers, and scientists as they were to the ancient hunters. This is not to say that we have made no progress over the decades to understand caribou population phenomena. Rather, our present understanding reveals a great deal of complexity that compels us to abandon the easy solutions offered by the universalistic frameworks of centralized resource management.

What might be the elements of a new approach to caribou science and management? First, we need to use a systems approach that considers people and caribou together. Second, we need adaptive management to grapple with the uncertainties of these social-ecological systems and to infuse them with a healthy dose of traditional knowledge. Third, resilience thinking helps to tackle complexity and to devise people-sensitive policies that can maintain long-term relationships between people and caribou. Fourth, learning and adaptation occur when management approaches are deliberated in co-management forums so that potential solutions reflect co-produced knowledge. I expand on each of these points briefly.

We need systems analysis to help integrate various factors and findings into a new synthesis that can explain both successes and failures, test hypotheses, and generate insights. Systems models have of course been used for many years in caribou population studies. But a new generation of systems analysis needs to consider the integrated caribou social-ecological system, including livelihood needs of people, hunters' motivations, Indigenous knowledge, the decision-making structure, and the governance of the resource at local and higher levels.

Adaptive management is learning-by-doing, and caribou hunters have carried out countless generations of what we would call adaptive management – the result being a rich culture of traditional ecological knowledge, or Indigenous knowledge. This knowledge is not merely about the natural history of the species; it also includes practices, beliefs, and values, which are embedded in stories, rituals of respect, and language. All of this knowledge is important for caribou management. The Indigenous wisdom clearly emphasizes uncertainties and the folly in trying to predict and control "the winds and the caribou."

Biologists may well want to internalize hunters' wisdom and manage caribou for resilience. That means, in brief, recognizing the reality of cycles; dampening the highs, protecting social and ecological memory at the lows, maintaining respectful relationships between people and caribou, building flexibility and options for the hunters, and not panicking when the caribou population counts give strange results. Indigenous knowledge holders have been advising us for a long time that caribou populations merge and split and shift calving areas. The herds are not as distinct as some think.

When uncertainty is high, the "expert knows best" approach is on very shaky ground. The smart manager minimizes risk by sharing it with the people whose livelihoods depend on the decisions. Stimulated by land claim agreements and the resulting co-management arrangements, researchers have been establishing relationships with Indigenous peoples as co-producers of locally relevant knowledge. These interactions create communities-of-learning that can pursue deliberate management goals and measures and that can co-produce knowledge for collaborative problem solving. Blueprint solutions do not work under conditions of uncertainty, and social learning is more likely to lead to adaptive responses.

This book contributes to a deeper understanding of the issues around caribou management. Many of the authors in the volume have been working on various aspects of caribou and caribou-people relationships. The chapters in this book show the importance of integrative, interdisciplinary approaches that include Indigenous voices. Caribou-dependent communities and resource managers need alternatives and backup options. Social learning and deliberation help to generate these options, building resilience in linked, interdependent, and co-evolving systems of caribou and people. Resilience and adaptation must be built on flexible, community-based management and participatory governance systems that enable us to learn from experience and to generate knowledge in order to cope with change.

Acknowledgments

The editors and authors are grateful for the leadership, support, and participation of the elders, harvesters, community researchers, and others from the Inuvialuit, Gwich'in, and Sahtú regions who were involved in the research and writing of this volume.

We particularly thank the staff and members of the Inuvialuit Game Council, Gwich'in Renewable Resources Board, and Sahtú Renewable Resources Board.

We express heartfelt gratitude for their support and contributions to Micheline Manseau, Deb Simmons, Tom Andrews, Jennifer Lam, Steve Baryluk, Jody Pellissey, Ray Case, Kristine Wray, and the editors and reviewers at UBC Press.

We gratefully acknowledge the financial support of the Social Sciences and Humanities Research Council of Canada, the Canadian International Polar Year Program, and the University of Alberta's Department of Resource Economics and Environmental Sociology in the Faculty of Agricultural, Life and Environmental Sciences, its Faculty of Native Studies, and its Department of Sociology in the Faculty of Arts.

When the Caribou Do Not Come

Introduction

Brenda Parlee and Ken Caine

The daily news headlines suggest we are witness to a surge in ecological crises around the world, many of which are having dramatic socio-economic and human health implications. In Canada alone, the loss of the Atlantic cod stocks, drought on the Prairies, flash flooding, and dramatic forest fire events have been triggers for rethinking our treatment of the environment as a pool of limitless resources. Many of these resource crises have been blamed, to some extent, on the failures of centralized, top-down, and rigid resource management approaches that have ignored the complexity and dynamics of ecosystems (Berkes, 2010; Holling, 2001; Holling & Meffe, 1996; Ludwig, Hilborn, & Walters, 1993). Essentially, "mother nature" has proven to be far more unpredictable and environmental problems far more intractable than anticipated. How can we cope with such unpredictability and ensure the sustainability of natural resources for future generations? Communities whose members have lived over many generations with ecological complexity and uncertainty may have some answers (Berkes, 2012; Howitt, 2001; Uphoff, 1998). In such communities, including Indigenous communities of northern Canada, well-developed systems of traditional knowledge have ensured the sustainability of natural resources and the well-being of communities over many generations (Berkes, Mathias, Kislalioglu, & Fast, 2001; Condon, Collings, & Wenzel, 1995; Freeman, Hudson, & Foote, 2005; Nuttall et al., 2005).

Northern Indigenous peoples who have had to deal with the ups and downs of barren-ground caribou populations are arguably among those with the greatest insights about how to cope with ecological complexity and uncertainty. Once numbering over 1.5 million in the Northwest Territories alone, barren-ground caribou have declined significantly over the past decade (Gunn, Russell, White, & Kofinas, 2009; Vors & Boyce, 2009). Why? Ecological data from the scientific community as well as traditional knowledge from many circumpolar nations detail historical oscillations in barren-ground caribou numbers on a cycle of forty to seventy years (Gunn, Russell, & Eamer, 2011; Gunn, Johnson, et al. 2011; Vors & Boyce, 2009). Despite such evidence, rapid declines in such an iconic and socio-economically important species have led to much tension and conflict in many parts of the Yukon, the Northwest Territories, Nunavut, and elsewhere in northern Canada. Many factors, such as expansive forest fires, weather events like the freezing-over of food sources, overgrazing on slow to regenerate tundra habitats, and climate change, are considered big picture drivers of population dynamics, with human disturbance, including resource development, being a critical concern to scientists and communities alike (Gunn, Johnson, et al., 2011; Johnson et al., 2005; Post & Forchhammer, 2002). However, it is subsistence harvesting by Indigenous peoples in the North that has been the preoccupation of many governments and publics. Why, despite little evidence of its impact, has Indigenous harvesting become almost the sole focus of wildlife management institutions in northern Canada in the past decade?

Those familiar with debates on the harvest of seals, whales, and polar bears might be quick to blame the animal rights movement. Seeking the protection of caribou and other iconic and charismatic species, environmental organizations have been prominent actors in the North and have arguably shaped a great deal of policy and debate related to wildlife conservation in recent years. Protecting the right to harvest amidst the interference of southern-based animal rights activists and organizations has been a challenge for Arctic Indigenous peoples since the beginning of the animal rights movement (Harter, 2004; Wenzel, 1991; Young, 1989).

But the answer is not that simple – there is much more going on within the North on questions of conservation than may be perceived from the outside. Indeed, from a social science perspective, the situation is also far more interesting than elsewhere – the devil is always in the details. Whereas some see the answer as more centralized governance and control (over Indigenous peoples), this position is in direct contrast to the perspectives of many northern Indigenous peoples, who perceive the problem to be too much centralized control.

Northern Indigenous peoples have been described and theorized as the original northern conservationists (Nadasdy, 2005). Since reports of declines in caribou populations, harvesters and communities have articulated much interest in engaging in discussion and creating voluntary harvest limits in order to do their part to ensure caribou populations are sustained for future generations (BCMPWG, 2011; PCMB, 2010). But for those with a clear eye on the past, including historians and Indigenous elders, the *imposition* of harvest limits may seem like history repeating itself. As early as 1894, which saw the creation of the Unorganized Territories Game Preservation Act, centralized governments started imposing harvest limits and criminalized many aspects of caribou-harvesting practices despite any evidence that harvesting was a factor in population declines, which occurred at the turn of the century and later in the 1950s (Campbell, 2004; Kulchyski & Tester, 2007; Ruttan, 2012; Sandlos, 2007; Usher, 2004). These limits, which seemingly had little ecological basis, were the cause of great social, economic, and cultural stress for communities already suffering from limited food resources and other impacts of colonialism, including the spread of European infectious diseases, residential school programs, and forced resettlement (Nadasdy, 1999, 2003; Piper & Sandlos, 2007; Regan, 2010).

On the whole, however, the drivers of population cycles, including the role of human disturbance, are still little understood, even for very well-studied subpopulations (Bergerud, 1996; Johnson & Russell, 2014). The quick assumption may be that more data are needed in order to better predict the timing and extent of population cycles. But barren-ground caribou systems are not linear or predictable; there are inherent uncertainties that are not entirely solvable or knowable. The critical issue according to many elders and leaders with a voice in this volume is not to predict or to "manage" the caribou but to respect them and deal with population dynamics in ways that ensure the sustainability of caribou and northern communities.

There are rich oral histories from many northern communities about the years of "so many caribou" and the years "when caribou did not come" – about what caribou meant for local cultures and identities, economies, the food on the table, as well as other aspects of their way of life. These oral histories, coupled with contemporary observations and experiences of decline and renewal, are the subject of this book. However, the research and narratives shared here are unique in voice and in temporal and spatial scale. Research underlying these chapters was carried out in collaboration with leaders and communities during a period of caribou population decline in a large area of northwestern Canada.

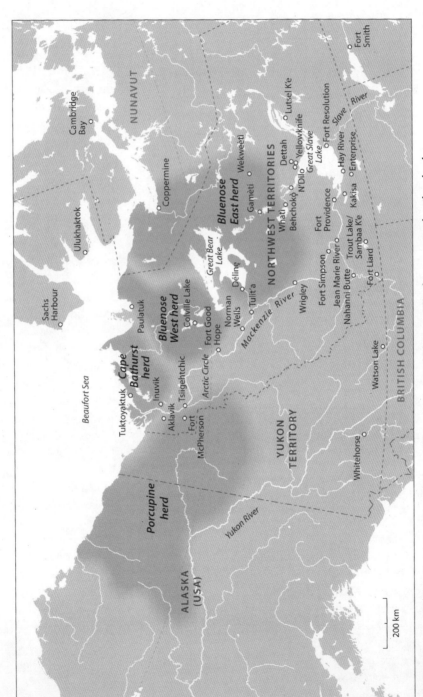

FIGURE I.I Ranges of the Porcupine, Cape Bathurst, Bluenose East, and Bluenose West barren-ground caribou herds

Much of the knowledge presented in this book may challenge readers more used to the conventional but negative stereotypes of northern Indigenous peoples as indiscriminate predators of caribou who have abandoned their traditional bow and arrow in favour of a more modern lifestyle (Collings, 1997; Sherry and Myers, 2002). This romanticized and rigid notion – that the cultures of Indigenous peoples should be frozen in time, circa 1899 – is highly problematic, if not racist, in its assumptions (Beavon, Voyageur, & Newhouse, 2005).

Although the livelihoods of northern Indigenous peoples have indeed changed over the past century, the core ways of life continue to mirror key aspects of the seasonal and year-to-year rhythms and cycles of their physical and spiritual worlds (Anderson & Nuttall, 2004; Ingold, 2000). This reality is very different from the one imaginable to the vast majority of Canadians. Although the North is part of our national identity, only a small number of Canadians have ever travelled north of the 60th parallel (Grace, 2002). Even fewer have experienced or seen a caribou, other than on the back of the Canadian twenty-five-cent coin.

Barren-ground caribou are known regionally as *tuktu* (Inuvialuitun), Ɂekwę (Sahtú), and *vadzaih* (Teetł'it Gwich'in). Biologists refer to the barren-ground caribou (*Rangifer tarandus groenlandicus* and *Rangifer tarandus granti*) of this region (the Tuktoyaktuk Peninsula being of limited discussion in this book) in terms of four main herds: Bluenose East, Bluenose West, Cape Bathurst, and Porcupine. These herds migrate thousands of kilometres through the taiga and boreal regions each year. Numbering over 500,000 in the 1990s, the population of these herds is now estimated to have fallen to half of that, a declining trend that parallels the declines in the other caribou herds in the Northwest Territories and elsewhere in northern Canada (Environment and Natural Resources, 2015; Vors & Boyce, 2009).

Given the symbolic importance of caribou to our national identity and the fundamental place of this species in the cultures and economies of northern Indigenous peoples, this book is likely to be of interest to northerners and the broader public alike. Although the work implicitly and explicitly tackles some complex theoretical problems of governance and stewardship, food security, and cultural continuity, not to mention the messy problem of caribou population dynamics, there is also something here for those simply interested in the mystique of the Arctic or alternative perspectives on wildlife management. There is no linear or prescriptive argument made by any of the authors; discussions about caribou population dynamics are complicated, as are the stories about how people deal with the dynamics of this iconic species.

Inuvialuit, Gwich'in, and Sahtú peoples, who have long histories of dependence and connection with caribou, recognize the inherent ecological variability associated with this species and have developed ways of understanding and coping with its socio-economic and cultural implications. These insights are not anecdotal opinions but are part of a system of knowledge, defined as traditional knowledge, that has developed over many generations. Traditional knowledge is the "cumulative body of knowledge, practice, and belief, evolving by adaptive processes and handed down through generations by cultural transmission, about the relationship of living beings (including humans) with one another and with their environment" (Berkes, Colding, & Folke, 2000, p. 1252).

The concept of traditional knowledge (also referred to as traditional ecological knowledge and Indigenous knowledge in this volume) is often framed or defined in the academic literature in relation to its potential integration with, or its contrast to, knowledge derived from Western science. Some scholars are quick to point out the synergies between the two knowledge systems – both of which rely heavily on systematic and empirical observation (Agrawal, 1995, 2002; Berkes, 2012; Roots, 1998). Although there is a deep spiritual dimension to the way that knowledge is generated and shared in many Indigenous cultures, there is also a tremendous empiricism that mirrors the rigour and systematic nature of the "scientific method."

Gwich'in, Sahtú, and Inuvialuit knowledge, like that of many other Indigenous cultures, is passed on through oral traditions, through shared observations, narratives, and songs (Blondin, 1990; Cruikshank, 1991), as well as through experiential practices of living on the land. Those traditional practices of common use include various kinds of ecological monitoring (systematic observation), temporally or spatially related harvest taboos of core species at different life stages, and habitat protection, to name a few (Berkes, 2012; Berkes, Colding, & Folke, 2000; Berkes & Turner, 2006; Gadgil, Berkes, & Folke, 1993; Moller et al., 2004; Parlee, Manseau, & Łutsël K'e Dene First Nation, 2005; Parlee et al., 2014).

The process of recognition and inclusion of traditional knowledge as a system of knowledge, practices, beliefs, and institutions (i.e., rules-in-use) is increasingly understood as a precursor or critical element of success in many resource management contexts, including forestry, agriculture, and coastal fisheries, as well as wildlife management (Berkes et al., 2001; Houde, 2007). Efforts to document traditional knowledge in northern Canada in ways that can influence or simply fit with existing decision-making processes are also growing; the methodology, technologies, and associated outcomes are diverse and include the documentation of detailed ethnographies, identification of archaeological sites, species inventories, ecological and cultural atlases, as well as knowledge archive systems, dictionaries, and geographic information systems (Auld & Kershaw, 2005; GRRB, 1997, 2001; Heine et al., 2001; Hennessy et al., 2013; Kritsch & Andre, 1997).

Within this emerging body of work are a growing number of new insights about northern ecosystems, including caribou ecology. Over the past several decades, regional government, co-management boards, academics, and research networks such as those funded by the International Polar Year have led to various kinds of initiatives featuring traditional knowledge in their research programs. Most notable is the work of the Arctic Borderlands Ecological Knowledge Co-op and related partner networks, such as the CircumArctic Rangifer Monitoring and Assessment Network, which has aimed to develop systematic and shared approaches to documenting both quantitative and qualitative observations about changes in the Porcupine caribou range (Eamer, 2006).

Although use of the term "traditional" may not be ideal to describe a knowledge system that is as contemporary as it is historical, in the Northwest Territories "traditional knowledge" is well integrated and accepted in legislation and land claim agreements, as well as used in myriad territorial and regional processes (e.g., environmental assessment and land-use planning) (Parlee, 2012). Co-management boards in the western Arctic have taken a lead role in facilitating the documentation and use of traditional knowledge in many forums, including wildlife co-management (Kendrick, 2003). However, the extent to which these efforts are valued and meaningful to scientists, community members, and policy makers varies significantly (Ellis, 2005; Nadasdy, 2003). Although there are numerous barriers and challenges, the requirement to include traditional knowledge in decision making presents unique and powerful opportunities (not yet realities) for many communities to be heard – opportunities enjoyed in few other jurisdictions in Canada and globally (Parlee, 2012).

A major challenge to meaningful inclusion of traditional knowledge is the recognition of "knowledge" as much more than anecdote, opinion, or simple data. Scholars and Indigenous community leaders involved in this volume offer a much deeper and broader understanding of knowledge as a system of learning that is well integrated with local beliefs and practices, as well as decision making at the individual, household, and community levels.

Acceptance of the notion that Indigenous peoples (also identified in other chapters as Aboriginal peoples) have both the knowledge and the capacity to manage their own resources, including wildlife, is difficult for some individuals and governments, particularly where old processes of decision making that privilege other values and uses of northern resources (e.g., mining) persist. These social and political processes of colonialism have been decades in the making and may take many decades more to reshape. In most parts of Canada, including the Northwest Territories, the legacy of more than a hundred years of colonization and a top-down governance system bent on transforming the North into a managed space (Piper & Sandlos, 2007) while "eradicating the Indian problem" (Scott, 1920) is still part of living memory and shapes identities and relationships of governance in the North, as it does elsewhere in Canada (Coulthard, 2007; Irlbacher-Fox, 2010). Understanding caribou management and the broader enterprise of natural resource management within this social and political landscape is crucial for understanding the issues and finding a meaningful way forward.

Does the current wildlife management landscape so mirror the past? Arguably, the devolution of power from federal centres to northern co-management boards has created tremendous positive change, with greater opportunities for community-based resource management than in many other places in Canada (Spaeder & Feit, 2005; Usher, 1995). Technology has also changed the knowledge being created and how it is used in decision making. The business of caribou management has become much more technical over the past fifty years as a result of the use of satellite data, aerial surveys, and computer modelling. For those involved in the development and use of such technology, there is increased precision and objectivity in the knowledge used in decision making; or has this technology simply obscured the value of lived experience and hidden the biases and subjectivities of data collection processes and management decisions in graphs and spreadsheets?

Using surveillance to predict the number of caribou in the four overlapping ranges of the Porcupine, Bluenose East, Bluenose West, and Cape Bathurst herds has not been without error and controversy. Early scholarship has told us that arithmetic counting of caribou is only one way of understanding population dynamics (Ruttan, 1966). Inuvialuit, Gwich'in, and Sahtú elders, leaders, and youth offer other kinds of accounting and explanations of when and why the "caribou do not come," presenting alternatives to the scientific models on calf recruitment and predation. Indigenous leaders and youth, including the authors in this book, remind us that the comings and goings of caribou are not fully knowable and that caribou have their own mind. As phrased by the ancient Dene in the saying that Fikret Berkes quotes in the Foreword to this book, "No one knows the way of the winds and the caribou" (Munsterhjelm, 1953, p. 97).

Given the extent of uncertainty and complexity associated with human-caribou relations, one might classify the task of caribou management as a "wicked problem" (Ludwig, 2001, p. 759). As noted by Chapin et al. (2008, p. 531), wicked problems are ones that "people disagree about how to define and solve ... in addition, efforts to solve the focal problem can create other secondary problems or unintended consequences."

To an outsider looking in or for those looking for simplistic answers, the situation may seem quite wicked or messy. But to those directly involved in caribou management over the past decade, the situation probably reads much differently. From either perspective, there may be cause for worry. A key concern emerging from this research is whether the tensions and conflicts over managing caribou that occurred within and between Indigenous communities, within the scientific community, and within government may have eroded the decades of trust building that have been at the core of the success of co-management institutions in the North (Kendrick, 2003). Or perhaps the conflicts that have surfaced over caribou management reveal some of the inequities of voice and power in the co-management processes that have been highlighted in other northern research (Howitt, 2001; Nadasdy, 2005).

The situation, however, is not entirely bleak and without solutions. According to oral traditions in some communities, if caribou are respected, they will come back to the people. Indeed, caribou numbers are beginning to recover in some areas (Environment and Natural Resources, 2015). At the outset of the research for this book in 2007, there was fear of a "collapse" of some herds, including the Porcupine herd, at the hands of Indigenous people, but that has turned out to be unfounded. For example, the Porcupine herd fell in numbers from a peak of 180,000 to 123,000 animals in 2001. At that time, Yukon newspaper headlines shrieked of immanent extirpation (Mostyn, 2010) and "orchestrated slaughters" by local Indigenous people who were hunting for their community (Thompson, 2008). However, in 2014 the population was reported to be at a new record high of nearly 200,000 animals, leading many to question how the scientists got it wrong. Equally concerning to Inuvialuit elder Frank Pokiak (pers. comm., 2015) was the fact that biologists did not address the error in any meaningful way or apologize to elders and the communities whose knowledge of herd dynamics had been ignored in previous debates and whose harvesting activities had been publicly critiqued as unsustainable.

Now is the perfect time to reflect on this period when the caribou did not come, while drawing on Indigenous knowledge to open a new conversation on this complex topic. Despite the complexity, there are guideposts for moving us forward. Caribou science has advanced significantly in recent years; although there are questions about the efficacy of some caribou counts, there is ever-greater awareness of the kinds of ecological variabilities characteristic of this species. There is also greater recognition of traditional knowledge in caribou management decision making.

We can look to the large body of previous traditional knowledge research to learn more about the depth and breadth of northern Indigenous peoples' knowledge of caribou ecology (Beaulieu, 2012; Ferguson & Messier, 1997; Gunn, Arlooktoo, & Kaomayok, 1988; Kendrick & Manseau, 2008; Kofinas, 2005; Kofinas et al., 2004; Legat, Chocolate, & Chocolate, 2008; Legat, Chocolate, Chocolate, et al., 2001; Legat, Chocolate, Gon, et al., 2001; Lyver, 2005; Parlee et al., 2014; Polfus et al., 2016; Thorpe, 1998; Zalatan, Gunn, & Henry, 2006; Zoe, 2012). There is also a corresponding and well-developed body of literature on the significance of caribou to cultural identities, economies, and health, which can be found in disciplines such as anthropology, cultural ecology, geography, political science, environmental history, and Indigenous/Native studies. More importantly, to learn more about the issues at hand, we can look to northern communities themselves and consider the new relationships that need to be forged and the lived experience that can be shared.

When the Caribou Do Not Come presents contributions and reflections from Aboriginal leaders, elders, and youth alongside the research of scholars in a range of social and natural science disciplines, including anthropology, rural sociology, political science, resource economics, history, environmental management, geography, and ecology. Together, we offer new perspectives on four key themes: counting caribou, understanding caribou, food security, and governance and management. The studies are primarily drawn from case study research in the Inuvialuit, Gwich'in, and Sahtú regions from 2007 to 2014, but they also draw on research across borders in the Yukon and Alaska. This body of work points to critical lessons about caribou and people as well as more theoretical and practical insights about how to deal with ecological complexity and uncertainty. Readers will find remarkable the extent to which communities have had to cope with a reported 70 percent decrease in a resource so fundamental to their culture, economies, and diets. Other scholars have explored such a capacity to deal with the dynamics of Arctic ecosystems in the fields of anthropology, geography, economics, and ecology (Anderson & Nuttall,

2004; Berkes, Colding, & Folke, 2000; Berman & Kofinas, 2004; Condon, Collings, & Wenzel, 1995; Forbes, 2008; Nuttall et al., 2005; Smith, 1978; Winterhalder, 1981). However, this capacity to cope with variability is increasingly complicated by the position of Indigenous people and the North within a growing global political economy; competing interests in natural resource development and natural resource conservation are altering the ways that caribou are valued and managed at the regional and national levels (Hummel & Ray, 2008). Climate change is also creating new kinds of patterns that are outside the scope of natural variability (Brotton & Wall, 1997; Krupnik & Jolly, 2002).

Communities in the Inuvialuit, Gwich'in, and Sahtú regions of the Northwest Territories have experienced previous periods of caribou population decline and have unique perspectives on the effects and appropriate responses to such dramatic ecological variability. Their experiences can provide useful lessons for those living within other dynamic ecosystems or for communities facing unprecedented changes in valued resources due to new pressures from resource development or climate change. Most importantly, they draw attention to the significance of community resilience.

SETTING THE STAGE

Northern Aboriginal peoples, including the Inuvialuit, Gwich'in, and Sahtú, have a cumulative body of knowledge, practice, belief, and institutions that has ensured the sustainability of northern resources for many generations. Such traditional knowledge is recognized in many kinds of legislation, notably three settled land claim agreements:

(1) Inuvialuit Final Agreement, 1984. This agreement created the Inuvialuit Settlement Region, which spans 906,430 square kilometres and includes several subregions: the Beaufort Sea, the Mackenzie River Delta, the northern portion of the Yukon (North Slope), the northwest portion of the Northwest Territories, and the western Canadian Arctic Islands. As part of the agreement, the Inuvialuit, territorial, and federal governments established the Joint Secretariat, a co-management arrangement that ensures representation of Inuvialuit in all aspects of wildlife, fisheries, and land and water management. The Inuvialuit Game Council oversees the management of game resources, including Bluenose West, Cape Bathurst, and Porcupine caribou.

(2) Gwich'in Comprehensive Land Claim Agreement, 1992. This agreement created the Gwich'in Settlement Area, which spans 56,935 square

kilometres and includes the communities of Aklavik, Fort McPherson, Inuvik, and Tsiigehtchic. As part of the agreement, the Gwich'in established the Gwich'in Renewable Resources Board as the main instrument of wildlife, fish, and forest management in the Gwich'in Settlement Area.

(3) Sahtú Dene and Métis Comprehensive Land Claim Agreement, 1994. This agreement created the Sahtú Settlement Area, which spans over 283,000 square kilometres and includes the communities in the Hare (K'asho Got'įnę), Great Bear Lake (Délįnę), and Mountain (Tulit'a) Districts. The Sahtú Renewable Resources Board, like its neighbouring board to the north, is the main instrument of wildlife and forestry management in the region. The aim is to assist communities with the management of wildlife and habitat for the benefit of the people of the Sahtú Settlement Area.

Among the land claim institutions with key roles in caribou management processes are the Inuvialuit Game Council, Gwich'in Renewable Resources Board, and Sahtú Renewable Resources Board. These organizations, in conjunction with local-level hunters and trappers associations and renewable resource councils, have played important roles in the research carried out for this book. The voices of their members are explicit and implicit in many aspects of the work presented here. Key leaders of each of these regional organizations are also authors of their own separate contributions, adding to the conversation on the book's major themes.

COUNTING CARIBOU

Much of the research for this book began with questions about reported declines in caribou numbers and the long-term sustainability of caribou herds in many parts of northern Canada. Among the most fundamental questions was "how are caribou declines defined?" Tensions in public hearings between communities and governments suggest that scientists and Indigenous peoples have different ways of accounting for population change. In Part 1, readers are challenged to consider the value of Inuvialuit, Gwich'in, and Sahtú oral histories as records of population decline, to discover how the historical experiences of northern peoples in more colonial periods of caribou management still matter, and to think about the subjectivities involved in different process of "counting caribou," including those associated with more technical methods. We present these alternative perspectives to stimulate discussion about the issues underlying the simplistic news headlines and discourse of a contemporary caribou crisis.

Much Indigenous knowledge about the comings and goings of caribou is grounded in systematic place-based observation of known caribou migration routes. Such observations are not recorded on digital spreadsheets but are chronicled in localized oral histories and passed on between harvesters and communities from generation to generation. Chapter 1, "From Tuktoyaktuk – Place of Caribou," provides a small glimpse into this unique place-based cultural record with Frank Pokiak's story about his first caribou hunt – the year the caribou came back to Tuktoyaktuk.

Indigenous oral histories, which are now considered valid legal evidence by Canadian supreme courts, have not been well respected in wildlife management decision making in previous years. In Chapter 2, "The Past Facing Forward," environmental historian John Sandlos uses archival records of the federal and territorial governments to explain more about the historical costs of ignoring such oral histories and the context for contemporary tensions in wildlife management. According to Sandlos, the present "caribou crisis" is not unlike previous periods of caribou population decline when Indigenous peoples and subsistence livelihoods were unjustly criminalized. Although much has changed since the settlement of comprehensive land claims, the contemporary period of caribou population decline suggests that deference still falls to scientists and scientific methods of counting caribou. Chapter 3, "Recounting Caribou," reveals some of the chinks in the armour of conventional scientific methods of counting caribou and tracking caribou movements. The chapter provides detail and examples that help to explain the dichotomy between science and traditional knowledge so often referenced in wildlife management and critiques of co-management. The term "recounting" has a double meaning here. It not only refers to the need for reflexivity in the production and use of population count data; it also refers to the need to recount, in the narrative and normative sense, how different worldviews, scales, and methods of observation can lead to very different outcomes and perspectives on Arctic ecosystem change.

Traditional knowledge is often limited in definition to a narrative and qualitative format, but there are other more quantitative methods in use by northern communities and governments that factor into knowledge systems, including those relevant to barren-ground caribou. In Chapter 4, "Beyond the Harvest Study," Brenda Parlee and colleagues discuss data collected through wildlife harvest studies in the Inuvialuit and Gwich'in regions. The reasons for wildlife harvest studies have been multifaceted; the Gwich'in and Inuvialuit harvest studies, like others in the Northwest Territories, Nunavut, and northern Quebec, were intended to establish a "minimum needs level" for the traditional harvest of country food. The data produced by counting harvests, including data on yield, location, and harvest effort, are also increasingly valued by biologists for the insights they can provide about wildlife ecology (Boyce, Baxter, & Possingham, 2012). Guided by theory and previous research on harvest as a proxy of population dynamics, the authors examine the Inuvialuit and Gwich'in harvest data, as well as reported caribou population estimates from the same period, to better understand how northern harvesters respond or adapt to changing ecological conditions, including the availability of barren-ground caribou. Whereas much discourse on northern harvesters assumes northern Indigenous peoples are opportunistic and indiscriminate predators, the analysis in Chapter 4 demonstrates a close synergy between resource availability and harvest.

UNDERSTANDING CARIBOU

The relationship of people to caribou is deeply spiritual. Many community elders in the Sahtú region, for example, describe themselves as Ɂekwęgot'ı̨nę, or "caribou people." This belief and sense of identity have been described and shared in previously documented oral histories by George Blondin and others (Blondin and Blondin, 2009). In Chapter 5, "We Are the People of the Caribou," photographer Morris Neyelle offers a photo essay through which readers can visually learn more about the practice of caribou hunting as well as the power of the caribou drum. His images and personal stories offer readers another way of understanding barren-ground caribou in the way of life of the community of Délı̨nę. In Chapter 6, "Harvesting in Dene Territory," Leon Andrew talks more about how these cultural identities and connections to caribou are important to the research process. As a Sahtú elder involved in various kinds of research initiatives over many years, he documents traditional knowledge about how to honour the caribou and his culture, providing insights for those interested in research outcomes and in respectful research relationships. But it is not only elders who have knowledge that is relevant to our understanding of the issues. In Chapter 7, "Dene Youth Perspectives," Roger McMillan discusses the issues facing Dene youth, specifically those of Fort Good Hope. How are they making sense of the changes occurring with caribou while coping with the many other kinds of environmental, socio-economic, and cultural changes they face in their personal lives and communities?

FOOD SECURITY

Caribou are an iconic species in many parts of the circumpolar North; in addition to having significant social, economic, and cultural significance, caribou meat is a major source of food in northern diets (Lambden, Receveur, & Kuhnlein, 2007; Usher, Duhaime, & Searles, 2003). The Government of the Northwest Territories has estimated the replacement value of caribou meat within community diets and economies to be in the millions of dollars per year. It is amidst this context of cultural and food security that we ask, "How do communities cope with declines in the availability of caribou?" In Chapter 8, "Time, Effort, Practice, and Patience," Anne Marie Jackson speaks to the importance of caribou as food for her family and community. As she is a young Sahtú Dene woman from Fort Good Hope, it is the teachings of her parents that provide a continuity or security through tough times when caribou are not around. She advises other youth to be patient and willing to learn and relearn the stories of previous generations and the associated practices; these stories will ensure that the Sahtú way of life is passed on to future generations. In Chapter 9, "The Wage Economy and Caribou Harvesting," the notion of food security is further discussed by Zoe Todd and Brenda Parlee, who explore the interconnected problems of caribou population decline, wage employment, and the regulation of caribou hunting. They ask how these problems factor into the food security of residents of Paulatuk in the Inuvailuit Settlement Region. The chapter challenges readers to consider how the dynamics of caribou population are not isolated from other aspects of environmental change (e.g., climate and impacts of mining) and socio-economic change in the community. The stress created for the household by the decrease in caribou meat availability can be compounded or offset by many other factors and influences, including industry actions and government decision making at the regional, territorial, and federal levels. The voices in this chapter remind us of the importance of managing food security, as Paulatukmiut do on a day-to-day basis, rather than taking a narrow or singular approach to coping with highly dynamic environmental and socio-economic stresses of life.

Among the most well-developed strategies for coping with variability in the availability of food resources has been food-sharing networks. Although many studies have focused on village-level food-sharing practices, in Chapter 10, "Caribou and the Politics of Sharing," Tobi Jeans Maracle and colleagues reveal how food sharing operates across both ecological and political boundaries. Their research, based in Old Crow, Yukon, tells the stories of those whose families, although living across the Northwest Territories and Alaska borders, are still part of the food-sharing networks that historically existed prior to the construction of such jurisdictions. In reading this chapter, we are challenged to think about the unique geographies of food security that exist in northern Canada and how those matter in the context of caribou management.

Governance and Management

A critical concern underlying this book is that the overemphasis on managing, if not controlling, Indigenous harvesting of caribou is a misdirected area of caribou management. History tells us that the "caribou crises" that are often misperceived during periods of population decline become social crises as a result of government efforts to limit or criminalize subsistence practices. The kinds of power that are exerted over caribou and over the caribou people are in fact due to conflicts over the value of the land and caribou to Indigenous cultures and economies as well as over their value to the public and others who view this charismatic species in a much more symbolic or Disney-like fashion. Although most people believe in the central notion of conservation, the role of hunting in conservation is less understood and accepted by governments and the public at large (Freeman, Hudson, & Foote, 2005; Nadasdy, 2007). A key problem seems to be the scope of conservation interest. Although many national and regional environmental organizations have typically been singular in their campaigns to "save the whale" (and more recently, the polar bear) or to "protect the old growth forests," these agendas have been exclusive of, and in some cases contrary to, the socio-economic and cultural dimensions of local livelihoods (Nadasdy, 2003). Conversely, notions of conservation among Indigenous communities are more complex; in addition to demonstrating an interest in conserving the environment and resources for future use, they stress the importance of the cultural and economic sustainability of their communities.

The systems of co-management currently in place in the North have made efforts to reconcile these disparate notions of conservation in their decision-making processes. As pointed out in earlier work on this theme, great strides have been made in addressing the disconnects of power and voice in decision making over wildlife through co-management, but there are still many gaps and challenges (Kendrick, 2003; Nadasdy, 2003).

What do northerners themselves think about this history and governance process? This book presents the voices of elders and leaders from three co-management boards, including Frank Pokiak of the Inuvialuit Game Council, Leon Andrew of the Sahtú Renewable Resources Board, and Robert Charlie of the Gwich'in Renewable Resources Board. These authors as well as the chapters by elder Morris Neyell and youth Anne Marie Jackson offer personal accounts of their own histories of caribou harvesting and some of the traditions of caribou management in their communities. In Chapter 11, "Recollections of Caribou Use and Management," Charlie provides this valuable perspective. Not only do readers catch a glimpse of the significance of caribou harvesting for Charlie and the harvesters of the Teetł'it Gwich'in, but they are also asked to think about management in a much different way. More than a formal boardroom process or set of guidelines and regulations enforced by outsiders, management is about the values and decisions of individual harvesters and their families. But how different are these "traditional rules" for managing caribou from those that are defined and legislated by co-management boards? In Chapter 12, "Ways We Respect Caribou," Kristine Wray explores this question based on research with the Teetł'it Gwich'in. In Chapter 13, "'Letting the Leaders Pass,'" Elisabeth Padilla and Gary P. Kofinas also highlight the ways that local rules for managing caribou matter by exploring some of the challenges in using traditional knowledge as the basis for formal regulations. In Chapter 14, "Linking the Kitchen Table and Boardroom Table," Brenda Parlee and colleagues offer ideas about the ways that decisions about caribou hunting made at the kitchen table matter at the boardroom table, reminding readers that formal governance systems, although largely comprised of men, need to be more considerate of women's perspectives. As the late Teetł'it Gwich'in elder Elizabeth Colin said, "Women have voice for the caribou too."

COMMUNITY RESILIENCE

Previous research has tackled the question of how individuals and societies cope with ecological variability and change; investigations have used different concepts and theories from social psychology, economic geography, cultural anthropology, and cultural ecology (Kofinas et al., 2010; Nuttall et al., 2005; Smith, 1978; Winterhalder, 2001). The concept of resilience is offered here as the lens through which we can explore the social dimensions of caribou population change. It recognizes the dynamic interrelationships between people and the environments in which they live.

The concept of resilience, although academic in nature and subject to critique (Davidson, 2010; Fabinyi, Evans, & Foale, 2014; Hornborg, 2009), is seen by some scholars as synergistic with the knowledge, practices, and worldviews of many Indigenous peoples, including those of northern Canada (Berkes, 2012). Theories on resilience that have emerged from the science of complex systems challenge conventional thinking about social responses to ecological change (Chapin, Folke, & Kofinas, 2009; Chapin et al., 2009; Ludwig, Hilborn, & Walters, 1993; Walker et al., 2004). Rather than attempting to control ecological complexities and uncertainties using linear, predictive, and disciplinary models, resilience thinkers focus on the importance of multiple ways of knowing and learning as the basis for adapting to ecosystem dynamics (Berkes et al., 2001; Chapin et al., 2006; Gunderson & Holling, 2002; Pahl-Wostl, 2009).

Resilience thinkers argue we must think and respond to ecological ups and downs in novel ways – being proactive and adaptive on an ongoing basis, not simply reactive to extreme situations. Within the context of a social-ecological system, the notion of resilience also directs us to explore the influence of the social, cultural, ecological, and political situation of institutions while paying attention to the unique social positions and ecological contexts of the various stakeholders involved (Angelstam et al., 2013). Doing so also requires thinking about the unique histories of particular regions and institutions rather than taking an ahistorical position and potentially repeating the mistakes of the past (Folke et al., 2007).

What do the chapters in this book tell us about resilience? What makes some communities and caribou systems more resilient than others? More research is needed to analyze multiple case studies in order to determine what core elements or conditions matter most and whether these vary across social, cultural, political, and ecological boundaries. Some preliminary insights based on the seven case studies in the Inuvialuit, Gwich'in, and Sahtú regions are offered here and may be generalizable, in some fashion, to other regions dealing with new kinds of ecological variability and change.

Accounts of previous exposure to, experiences with, and traditional knowledge about the ecological problem – caribou population dynamics – are among the fundamental offerings of *When the Caribou Do Not Come*. Those elders with oral histories about the population cycles of barren-ground caribou have much to teach us about how to cope with the most recent decline. But not all individuals and communities have the same kinds of oral histories due to differing cultural practices and norms for sharing traditional knowledge, different geographic locations within a caribou range, or other socio-cultural factors. The protection of intellectual property rights has been a pervasive concern for many Indigenous communities in northern Canada and elsewhere. Rules for dealing with the question of intellectual property rights in ways that enable communities to continue building, using, and sharing their knowledge seem more well developed where land and resource rights are clearly protected, as they are in the Inuvialuit, Gwich'in, and Sahtú regions. For example, the Gwich'in Tribal Council through the Gwich'in Social and Cultural Institute has its own traditional knowledge policy, which dictates how, where, and to whom knowledge can be reported. The security offered through this institution has enabled great strides to be made in the documentation and use of Gwich'in knowledge in this region. In other jurisdictions, where such security does not exist, the sharing and use of traditional knowledge can be more challenging both inside and outside communities.

The fostering of diverse livelihood options enables communities to better ride out periods when the caribou do not come; some communities have less access to other kinds of food resources, including traditional/country foods, which makes diversification difficult. Sudden or abrupt efforts to substitute one food resource for another or to make livelihood shifts during times of stress are more difficult than in communities where livelihood diversification has been nurtured over time. Part of that story hinges on the availability of knowledge and on the capacity to access these diverse resources or to engage in more diverse kinds of livelihood practices. For communities whose cultural traditions, including traditional knowledge, have been lost or eroded due to internal or external cultural pressures (e.g., residential school programs), the rediscovery of oral histories about the past and the reinterpretation or recasting of that knowledge in new contexts can contribute to the capacity to cope with new challenges (Napoleon, 2013).

In addition to oral histories, there also appears to be a necessity for ongoing tracking and communication between harvesters and harvester communities in the range. This idea of tracking or monitoring is not new to Indigenous communities in the Inuvialuit, Gwich'in, and Sahtú regions. Initiatives such as the Arctic Ecological Borderlands Knowledge Co-op and other processes of community-based monitoring that honour community observations and community-generated "data" about changing ecosystems can also support resilience. But sharing is the key; collecting data about ecological conditions for spreadsheets alone does not contribute to the social learning considered to be so important to resilience.

The role of the community in formal governance, including co-management systems, also matters significantly. In situations where Indigenous harvesters have power in the decision-making process, there is much more opportunity for ensuring tight feedbacks (or strong links) between the traditional knowledge that is being generated and management outcomes. However, the key issue is flexibility; institutions that are rigid in using a one-size-fits-all and top-down approach have served to undermine rather than empower communities in dealing with the day-to-day realities of caribou management that go beyond the boardroom table. A case in point is the situation of the Yukon government, which made efforts to override a co-managed process of caribou harvest management being developed for the Porcupine caribou herd and was met with much resistance, including legal action (CBC, 2010). This government intervention seemed to be a step backward in the harvest management planning process. What does this example tell us? Perhaps that the imposition of top-down regulations across communities that have their own set of rules for caribou management based on generations of traditional knowledge (as discussed in Chapters 12 and 13) serves only to weaken the relationships and learning among various stakeholders considered necessary for the resilience of communities.

The continued generation and use of traditional knowledge are also an underlying thread in our understanding of resilience; as suggested by Jackson in Chapter 8, the stories and teachings of the elders provide the support or continuity for the community to go through hard times. The continued ability of youth and others to live and sustain their families on the land is the fundamental issue that should be of concern in caribou management, as discussed by the late Elizabeth Colin (see Chapter 14). Other kinds of stresses also factor into the equation of resilience. Readers are challenged to think about whether climate change, resource development, and the imposition of centralized rules and regulations decrease the capacity of communities to learn, cope, and adapt or whether these influences support and foster resilience.

Resilience is also about relationships and ongoing dialogue. The editors of this volume, although previous residents of the Northwest Territories, no longer live north of the 60th parallel. We are aware of how southerners (including academics) are perceived by northerners, so we are persistently reflexive in our interpretation of the value of the volume. Too often, those from southern Canada are convinced they must bring ideas, technologies, and resources north to address northern problems. We are hopeful that the chapters in this volume will stimulate discussion and conversation but mostly that readers will be compelled to listen to those who have other stories to tell about "when the caribou do not come."

References

Angelstam, P., Andersson, K., Annerstedt, M., Axelsson, R., Elbakidze, M., Garrido, P., . . ., & Stjernquist, I. (2013). Solving problems in social-ecological systems: Definition, practice and barriers of transdisciplinary research. *Ambio, 42*(2), 254–265.

Agrawal, A. (1995). Dismantling the divide between indigenous and scientific knowledge. *Development and Change, 26*(3), 413–439.

Agrawal, A. (2002). Indigenous knowledge and the politics of classification. *International Social Science Journal, 54*(173), 287–297.

Anderson, D.G. & Nuttall, M. (Eds.). (2004). *Cultivating Arctic Landscapes: Knowing and Managing Animals in the Circumpolar North.* New York: Berghahn Books.

Auld, J., & Kershaw, R. (Eds.) (2005). *The Sahtu Atlas: Maps and Stories from the Sahtu Settlement Area in Canada's Northwest Territories.* Norman Wells, NWT: Sahtu GIS Project.

Beaulieu, D. (2012). Dene traditional knowledge about caribou cycles in the Northwest Territories. *Rangifer, 32*(2), 59–67.

Beavon, D.J., Voyageur, C.J., & Newhouse, D.R. (2005). *Hidden in Plain Sight: Contributions of Aboriginal Peoples to Canadian Identity and Culture.* Toronto: University of Toronto Press.

Bergerud, A.T. (1996). Evolving perspectives on caribou population dynamics, have we got it right yet? *Rangifer, 16*(9), 95–116.

Berkes, F. (2010). Devolution of environment and resources governance: Trends and future. *Environmental Conservation, 37*(4), 489–500.

Berkes, F. (2012). *Sacred Ecology* (3rd ed.). New York: Routledge.

Berkes, F., Colding, J., & Folke, C. (2000). Rediscovery of traditional ecological knowledge as adaptive management. *Ecological Applications, 10*(5), 1251–1262.

Berkes, F., Mathias, J., Kislalioglu, M., & Fast, H. (2001). The Canadian Arctic and the Oceans Act: The development of participatory environmental research and management. *Ocean and Coastal Management, 44*(7–8), 451–69. https://doi.org/10.1016/s0964 -5691(01)00060-6

Berkes, F., & Turner, N.J. (2006). Knowledge, learning and the evolution of conservation practice for social-ecological system resilience. *Human Ecology, 34*(4), 479–494.

Berman, M., & Kofinas, G. (2004). Hunting for models: Grounded and rational choice approaches to analyzing climate effects on subsistence hunting in an Arctic community. *Ecological Economics, 49*(1), 31–46.

Blondin, G. (1990). *When the World Was New: Stories of the Sahtú Dene.* Yellowknife: Outcrop, the Northern Publisher.

Blondin, G., & Blondin, J. (2009). *The Legend of the Caribou Boy*. Penticton, BC: Theytus.

Bluenose Caribou Management Plan Working Group (BCMPWG) (2011). *Taking Care of Caribou: Cape Bathurst, Bluenose-West, and Bluenose-East Barren-Ground Caribou Herds Management Plan*. Terriplan Consultants (Ed.). Yellowknife: Department of Environment and Natural Resources, Government of the Northwest Territories.

Boyce, M.S., Baxter, P.W., & Possingham, H.P. (2012). Managing moose harvests by the seat of your pants. *Theoretical Population Biology, 82*(4), 340–347.

Brotton, J., & Wall, G. (1997). Climate change and the Bathurst caribou herd in the Northwest Territories. *Climatic Change, 35*(1), 35–52.

Campbell, C. (2004). A genealogy of the concept of "wanton slaughter" in Canadian wildlife biology. In D.G. Anderson & M. Nutall (Eds.), *Cultivating Arctic Landscapes: Knowing and Managing Animals in the Circumpolar North* (pp. 154–171). New York: Berghahn Books.

CBC (2010). N.W.T. groups take Yukon to court over caribou hunting rules. *CBC News,* February 2. http://www.cbc.ca/news/canada/north/n-w-t-groups-take-yukon -to-court-over-caribou-hunting-rules-1.966412.

Chapin, F.S., Folke, C., & Kofinas, G. (2009). A framework for understanding change. In F.S. Chapin, G. Kofinas, & C. Folke (Eds.), *Principles for Ecosystem Stewardship: Resilience-Based Natural Resource Management in a Changing World* (pp. 3–28). New York: Springer.

Chapin, F.S., Hoel, M., Carpenter, S.R., Lubchenco, J., Walker, B., Callaghan, T.V., . . ., & Zimov, S.A. (2006). Building resilience and adaptation to manage Arctic change. *Ambio, 35*(4), 198–202.

Chapin, F.S., Kofinas, G., Folke, C., Carpenter, S.R., Olsson, P., Abel, N., . . ., & Young, O.R. (2009). Resilience-based stewardship: Strategies for navigating sustainable pathways in a changing world. In F.S. Chapin, G. Kofinas, & C. Folke (Eds.), *Principles for Ecosystem Stewardship: Resilience-Based Natural Resource Management in a Changing World* (pp. 319–337). New York: Springer.

Chapin, F.S., Trainor, S.F., Huntington, O., Lovecraft, A.L., Zavaleta, E., Natcher, D.C., . . ., & Naylor, R.L. (2008). Increasing wildfire in Alaska's boreal forest: Pathways to potential solutions of a wicked problem. *Bioscience, 58*(6), 531–40.

Collings, P. (1997). The cultural context of wildlife management in the Canadian North. In E.A. Smith & J. McCarter (Eds.), *A Contested Arctic: Indigenous People, Industrial States, and the Circumpolar Environment* (pp. 13–40). Seattle: University of Washington Press.

Condon, R., Collings, P., & Wenzel, G. (1995). The best part of life: Subsistence hunting, ethnicity, and economic adaptation among young adult Inuit males. *Arctic, 48*(1), 31–46.

Coulthard, G.S. (2007). Subjects of empire: Indigenous peoples and the "politics of recognition" in Canada. *Contemporary Political Theory, 6*(4), 437–460.

Cruikshank, J. (1991). *Life Lived Like a Story: Life Stories of Three Yukon Native Elders*. Lincoln: University of Nebraska Press.

Davidson, D. (2010). The applicability of the concept of resilience to social systems: Some sources of optimism and nagging doubts. *Society & Natural Resources, 23*(12), 1135–1149.

Eamer, J. (2006). Keep it simple and be relevant: The first ten years of the Arctic Borderlands Ecological Knowledge Co-op. In W.V. Reid, F. Berkes, T. Wilbanks, &

D. Capistrano (Eds.), *Bridging Scales and Knowledge Systems* (pp. 185–206). Washington, DC: Island.

Ellis, S.C. (2005). Meaningful consideration? A review of traditional knowledge in environmental decision making. *Arctic, 58*(1), 66–77.

Environment and Natural Resources (2015). Trends in barren-ground caribou population size in tundra-taiga ecosystems. http://www.enr.gov.nt.ca/state-environment/154-trends -barren-ground-caribou-population-size-tundra-taiga-ecosystems.

Fabinyi, M., Evans, L., & Foale, S.J. (2014). Social-ecological systems, social diversity, and power: Insights from anthropology and political ecology. *Ecology and Society, 19*(4), 28.

Ferguson, M.A.D., & Messier, F. (1997). Collection and analysis of traditional ecological knowledge about a population of Arctic tundra caribou. *Arctic, 50*(1), 17–28.

Folke, C., Pritchard, L., Jr., Berkes, F., Colding, J., & Svedin, U. (2007). The problem of fit between ecosystems and institutions: Ten years later. *Ecology and Society, 12*(1), 30.

Forbes, B.C. (2008). Equity, vulnerability and resilience in social-ecological systems: A contemporary example from the Russian Arctic. *Research in Social Problems and Public Policy, 15*, 203–236.

Freeman, M., Hudson, R., & Foote, L. (2005). *Conservation Hunting: People and Wildlife in Canada's North*. Edmonton: Canadian Circumpolar Institute Press.

Gadgil, M., Berkes, F., & Folke, C. (1993). Indigenous knowledge for biodiversity conservation. *Ambio, 22*(2–3), 151–156.

Grace, S.E. (2002). *Canada and the Idea of North*. Montreal, Kingston: McGill-Queen's University Press.

Gunderson, L.H., & Holling, C.S. (2002). *Panarchy: Understanding Transformations in Human and Natural Systems*. Washington, DC: Island.

Gunn, A., Johnson, C., Nishi, J., Daniel, C., Russell, D.E., Carlson, M., & Adamczewski, J. (2011). Understanding the cumulative effects of human activities on barrenground caribou. In P.R. Krausman & L.K. Harris (Eds.), *Cumulative Effects in Wildlife Management: Impact Mitigation* (pp. 113–133). Boca Raton, FL: CRC Press.

Gunn, A., Russell, D., & Eamer, J. (2011). *Northern Caribou Population Trends in Canada*. Ottawa: Canadian Councils of Resource Ministers.

Gunn, A., Russell, D., White, R., & Kofinas, G. (2009). Facing a future of change: Wild migratory caribou and reindeer. *Arctic, 62*(3), iii–iv.

Gunn, A., Arlooktoo, G., & Kaomayok, D. (1988). The contribution of the ecological knowledge of Inuit to wildlife management in the Northwest Territories. In M.M.R. Freeman & L.N. Carbyn (Eds.), *Traditional Knowledge and Renewable Resource Management in Northern Regions* (pp. 22–30). Edmonton: IUCN Commission on Ecology and the Boreal Institute for Northern Studies.

Gwich'in Renewable Resources Board (GRRB). (1997). *Nành' Kak Geenjit Gwich'in Ginjik (Gwich'in Words about the Land)*. Inuvik: Gwich'in Renewable Resources Board.

Gwich'in Renewable Resources Board (GRRB). (2001). *Gwìndòo Nành' Kak Geenjit Gwich'in Ginjik (More Gwich'in Words about the Land)*. Inuvik: Gwich'in Renewable Resources Board.

Harter, J.-H. (2004). Environmental justice for whom? Class, new social movements, and the environment: A case study of Greenpeace Canada, 1971–2000. *Labour (Halifax), 54*, 83–119.

Heine, M., Andre, A., Kritsch, I., & Cardinal, A. (2001). *Gwichya Gwich'in Googwandak: The History and Stories of the Gwichya Gwich'in as Told by the Elders of Tsiigehtchic*. Tsiigehtchic and Yellowknife: Gwich'in Social and Cultural Institute.

Hennessy, K., Lyons, N., Loring, S., Arnold, C., Joe, M., Elias, A., & Pokiak, J. (2013). The Inuvialuit living history project: Digital return as the forging of relationships between institutions, people, and data. *Museum Anthropology Review*, 7(1/2), 44–73.

Holling, C.S. (2001). Understanding the complexity of economic, ecological, and social systems. *Ecosystems (New York, NY)*, 4(5), 390–405.

Holling, C.S., & Meffe, G.K. (1996). Command and control and the pathology of natural resource management. *Conservation Biology*, 10(2), 328–337.

Hornborg, A. (2009). Zero-sum world: Challenges in conceptualizing environmental load displacement and ecologically unequal exchange in the world-system. *International Journal of Comparative Sociology*, 50(3–4), 237–62.

Houde, N. (2007). The six faces of traditional ecological knowledge: Challenges and opportunities for Canadian co-management arrangements. *Ecology and Society*, 12(2), 34.

Howitt, R. (2001). *Rethinking Resource Management: Justice, Sustainability and Indigenous Peoples*. London, New York: Routledge.

Hummel, M., & Ray, J. (2008). *Caribou and the North: A Shared Future*. Toronto: Dundurn.

Ingold, T. (2000). *Perceptions of the Environment: Essays in Livelihood, Dwelling and Skill*. New York: Routledge.

Irlbacher-Fox, S. (2010). *Finding Dahshaa: Self-Government, Social Suffering, and Aboriginal Policy in Canada*. Vancouver: UBC Press.

Johnson, C.J., Boyce, M.S., Case, R.L., Cluff, H.D., Gau, R.J., Gunn, A., & Mulders, R. (2005). Cumulative effects of human developments on Arctic wildlife. *Wildlife Monographs*, 160(1), 1–36.

Johnson, C.J., & Russell, D.E. (2014). Long-term distribution responses of a migratory caribou herd to human disturbance. *Biological Conservation*, 177, 52–63.

Kendrick, A. (2003). The flux of trust: Caribou co-management in northern Canada. *Environments*, 31(1), 43–59.

Kendrick, A., & Manseau, M. (2008). Representing traditional knowledge: Resource management and Inuit knowledge of barren-ground caribou. *Society & Natural Resources*, 21(5), 404–418.

Kofinas, G. (2005). Caribou hunters and researchers at the co-management interface: Emergent dilemmas and the dynamics of legitimacy in power sharing. *Anthropologica*, 47(2), 179–96.

Kofinas, G., Chapin, F.S., BurnSilver, S., Schmidt, J.I., Fresco, N.L., Kielland, K., Martin, S., Springsteen, A., & Rupp, T.S. (2010). Resilience of Athabascan subsistence systems to interior Alaska's changing climate. *Journal of Forest Research*, 40(7), 1347–1359.

Kofinas, G., Lyver, P.O., Russell, D., White, R., Nelson, A., & Flanders, N. (2004). Towards a protocol for monitoring of caribou body condition. *Rangifer*, (special issue 14), 43–52.

Kritsch, I.D., & Andre, A.M. (1997). Gwich'in traditional knowledge and heritage studies in the Gwich'in Settlement Area. In G.P. Nicholas & T.D. Andrews (Eds.), *At a Crossroads: Archaeology and First Peoples in Canada* (pp. 125–144). Burnaby, BC: Archaeology Press, Simon Fraser University.

Krupnik, I., & Jolly, D. (2002). *The Earth Is Faster Now: Indigenous Observations of Arctic Environmental Change*. Fairbanks, AK: Arctic Research Consortium of the United States.

Kulchyski, P., & Tester, F. (2007). *Kiamujut (Talking Back): Game Management and Inuit Rights, 1900–70*. Vancouver: UBC Press.

Lambden, J., Receveur, O., & Kuhnlein, H.V. (2007). Traditional food attributes must be included in studies of food security in the Canadian Arctic. *International Journal of Circumpolar Health, 66*(4), 308–319.

Legat, A., Chocolate, G., & Chocolate, M. (2008). *Monitoring the Relationship between People and Caribou: Tłįchǫ Laws and Indicators of Change*. Yellowknife: West Kitikmeot Slave Study Society.

Legat, A., Chocolate, G., Chocolate, M., Williah, P., & Zoe, S.A. (2001). *Habitat of Dogrib Traditional Territory: Placenames as Indicators of Biogeographical Knowledge*. Yellowknife: West Kitikmeot Slave Study Society.

Legat, A., Chocolate, G., Gon, B., Zoe, S.A., & Chocolate, M. (2001). *Relationship between Caribou Migration Patterns and the State of Caribou Habitat*. Yellowknife: West Kitikmeot Slave Study Society.

Ludwig, D. (2001). The era of management is over. *Ecosystems (New York, NY), 4*(8), 758–64.

Ludwig, D., Hilborn, R., & Walters, C. (1993). Uncertainty, resource exploitation, and conservation: Lessons from history. *Science*, (April), 2, 17–36.

Lyver, P.O. (2005). Monitoring barren-ground caribou body condition with Denésǫłıné traditional knowledge. *Arctic, 58*(1), 44–54.

Moller, H., Berkes, F., Lyver, P.O., & Kislalioglu, M. (2004). Combining science and traditional ecological knowledge: Monitoring populations for co-management. *Ecology and Society, 9*(3), 2.

Mostyn, R. (2010). Say goodbye to the caribou. *Yukon News,* February 19. http://www.yukon-news.com/letters-opinions/say-goodbye-to-the-caribou.

Munsterhjelm, E. (1953). *The Wind and the Caribou: Hunting and Trapping in Northern Canada*. London: George Allen and Unwin.

Nadasdy, P. (1999). The politics of TEK: Power and the "integration of knowledge." *Arctic Anthropology, 36*(1–2), 1–18.

Nadasdy, P. (2003). *Hunters and Bureaucrats: Power, Knowledge and Aboriginal State Relations in the Southwest Yukon*. Vancouver: UBC Press.

Nadasdy, P. (2005). Transcending the debate over the ecologically noble Indian: Indigenous peoples and environmentalism. *Ethnohistory (Columbus, OH), 52*(2), 291–331.

Nadasdy, P. (2007). The gift in the animal: The ontology of hunting and human-animal sociality. *American Ethnologist, 34*(1), 25–43.

Napoleon, V. (2013). Thinking about Indigenous legal orders. In R. Provost & C. Sheppard (Eds.), *Dialogues on Human Rights and Legal Pluralism* (pp. 229–245). New York: Springer.

Nuttall, M., Berkes, F., Forbes, B., Kofinas, G., Vlassova, T., & Wenzel, G. (2005). Hunting, herding, fishing, and gathering: Indigenous peoples and renewable resource use in the Arctic. In C. Symon, L. Arris, & B. Heal (Eds.), *Arctic Climate Impact Assessment* (pp. 649–690). New York: Cambridge University Press.

Pahl-Wostl, C. (2009). A conceptual framework for analysing adaptive capacity and multi-level learning processes in resource governance regimes. *Global Environmental Change, 19*(3), 354–365.

Parlee, B. (2012). Finding voice in a changing ecological and political landscape: Traditional knowledge and resource management in settled and unsettled land claim areas of the Northwest Territories, Canada. *Aboriginal Policy Studies, 2*(1), 56–87.

Parlee, B., Goddard, E., Basil, M., & Smith, M. (2014). Tracking change: Traditional knowledge of wildlife health in northern Canada. *Human Dimensions of Wildlife, 19*(1), 47–61.

Parlee, B., Manseau, M., & Łutsël K'e Dene First Nation. (2005). Using traditional knowledge to adapt to ecological change: Denésǫłıné monitoring of caribou movements. *Arctic, 58*(1), 26–37.

Piper, L., & Sandlos, J. (2007). A broken frontier: Ecological imperialism in the Canadian North. *Environmental History, 12*(4), 759–795.

Polfus, J., Manseau, M., Simmons, D., Neyelle, M., Bayha, W., Andrew, F., & Wilson, P. (2016). Łeghágots' enetę (learning together): The importance of Indigenous perspectives in the identification of biological variation. *Ecology and Society, 21*(2), 18.

Porcupine Caribou Management Board (PCMB). (2010). *Harvest Management Plan for the Porcupine Caribou Herd in Canada.* Yellowknife: Na-Cho Nyak Dun First Nation, Gwich'in Tribal Council, Inuvialuit Game Council, Tr'ondëk Hwëch'in First Nation, Government of Vuntut Gwitchin, Government of the Northwest Territories, Government of the Yukon Territory, and Government of Canada.

Post, E., & Forchhammer, M.C. (2002). Synchronization of animal population dynamics by large-scale climate. *Nature,* (November): 14, 168–171.

Regan, P. (2010). *Unsettling the Settler Within: Indian Residential Schools, Truth Telling, and Reconciliation in Canada.* Vancouver: UBC Press.

Roots, F. (1998). Inclusion of different knowledge systems in research. In M. Manseau (Ed.), *Terra Borealis 1: Traditional and Western Scientific Environmental Knowledge* (pp. 42–49). Goose Bay, NL: Institute for Environmental Monitoring and Research.

Ruttan, R.A. (1966). New crisis for barren-ground caribou. *Country Guide,* September. Library and Archives Canada, RG 109, vol. 381, file 24, WLU 200.

Ruttan, R.A. (2012). New caribou crisis – Then and now. *Rangifer, 32*(20), 85–102.

Sandlos, J. (2007). *Hunters on the Margin: Native People and Wildlife Conservation in the Northwest Territories.* Vancouver: UBC Press.

Scott, D.C. (1920). The Indian problem. Library and Archives Canada, RG 10, vol. 6810, file 470-2-3, vol. 7, pp. 55 (L-3) and 63 (N-3).

Sherry, E., & Myers, H. (2002). Traditional environmental knowledge in practice. *Society & Natural Resources, 15*(4), 345–358.

Smith, J.G.E. (1978). Economic uncertainty in an "original affluent society": Caribou and caribou eater Chipewyan adaptive strategies. *Arctic Anthropology, 15*(1), 68–88.

Spaeder, J., & Feit, H.A. (2005). Co-management and Indigenous communities: Barriers and bridges to decentralized resource management. *Anthropologica, 47*(2), 147–154.

Thompson, J. (2008). Orchestrated slaughter threatens Porcupine caribou. *Yukon News,* September 6. http://yukon-news.com/news/orchestrated-slaughter-threatens-porcupine-caribou.

Thorpe, N.L. (1998). The Hiukitak School of Tuktu: Collecting Inuit ecological knowledge of caribou and calving areas through an elder-youth camp. *Arctic, 51*(4), 403–408.

Uphoff, N. (1998). Community-based natural resource management: Connecting micro and macro processes and people with their environments. Paper presented at the International Workshop on Community-Based Natural Resource Management, Washington, DC.

Usher, P.J. (1995). Co-management of natural resources: Some aspects of the Canadian experience. In D.L. Peterson & D.R. Johnson (Eds.), *Human Ecology and Climate Change: People and Resources in the Far North* (pp. 197–206). Washington, DC: Taylor and Francis.

Usher, P.J. (2004). Caribou crisis or administrative crisis? Wildlife and Aboriginal policies on the barren grounds of Canada, 1947–60. In D.G. Anderson & M. Nutall (Eds.), *Cultivating Arctic Landscapes: Knowing and Managing Animals in the Circumpolar North* (pp. 172–199). New York: Berghahn Books.

Usher, P.J., Duhaime, G., & Searles, E. (2003). The household as an economic unit in Arctic Aboriginal communities, and its measurement by means of a comprehensive survey. *Social Indicators Research, 61*(2), 175–202. https://doi.org/10.1023/a:1021344707027

Vors, L.S., & Boyce, M.S. (2009). Global declines of caribou and reindeer. *Global Change Biology, 15*(11), 2626–2633. https://doi.org/10.1111/j.1365-2486.2009.01974

Walker, B., Holling, C.S., Carpenter, S., & Kinzig, A. (2004). Resilience, adaptability and transformability in social-ecological systems. *Ecology and Society, 9*(2), 5.

Wenzel, G.W. (1991). *Animal Rights, Human Rights: Ecology, Economy, and Ideology in the Canadian Arctic.* Toronto: University of Toronto Press.

Winterhalder, B. (1981). Optimal foraging strategies and hunter-gatherer research in anthropology: Theories and models. In B. Winterhalder & E.A. Smith (Eds.), *Hunter-Gatherer Foraging Strategies: Ethnographic and Archeological Analyses* (pp. 13–35). Chicago: University of Chicago Press.

Winterhalder, B. (2001). The behavioral ecology of hunter-gatherers. In C. Panter-Brick, R.H. Layton, & P. Rowley-Conwy (Eds.), *Hunter-Gatherers: An Interdisciplinary Perspective* (pp. 12–38). Cambridge, UK: Cambridge University Press.

Young, O.R. (1989). The politics of animal rights: Preservationists vs. consumptive users in the North. *Études/Inuit/Studies, 13*(1), 43–59.

Zalatan, R., Gunn, A., & Henry, G. (2006). Long-term abundance patterns of barren-ground caribou using trampling scars on roots of *Picea mariana* in the Northwest Territories, Canada. *Arctic, Antarctic, and Alpine Research, 38*(4), 624–630.

Zoe, J.B. (2012). Ekwǫ̀ and Tłįchǫ nàowo/Caribou and Tłįchǫ language, culture and way of life: An evolving relationship and shared history. *Rangifer, 32*(2), 69–74.

PART 1
Counting Caribou

I

From Tuktoyaktuk – Place of Caribou

Frank Pokiak

The first time I went hunting caribou was in 1967. My brother-in-law sent for us boys to go to Aklavik to hunt caribou with him. He said that there were a lot of caribou in the foothills not too far from Aklavik. When we went out hunting, we went eight kilometres upriver from Aklavik. We travelled from there for about one and a half hours and eventually ran into a herd of about 150 caribou. We got six caribou that trip. That was the very first time I hunted caribou. I stayed with my brother-in-law and his family for about four years. Sometimes we travelled far away to find caribou – the farthest I had to go hunting was above Canoe Lake around Cache Creek and Fish Creek. At that time, a group of us would go, and we would share all the meat equally among us. Like many other Inuvialuit, I have seen the caribou come and go. The name of my home, Tuktoyaktuk, means "a place where there is caribou."* This coming and going of the caribou is a natural cycle and one that our traditional Inuvialuit culture accepts as part of our lives.

There was a time when there were many caribou around Tuktoyaktuk. Then some time in the late 1920s, the caribou left the area. The government at the time believed that without caribou, the Inuvialuit would experience hardships, so they brought in the reindeer herd from Alaska

---

* Tuktoyaktuk is at the northern end of a peninsula extending into Kugmallit Bay, just east of the Mackenzie River Delta, on the Arctic Coast. It was formerly called Port Brabant, and the present name is commonly abbreviated to "Tuk." Tuktoyaktuk is 122 kilometres by air or 177 kilometres by river north of Inuvik.

in the 1930s. It took them five years to bring the herd from Alaska to the Mackenzie Delta area. The herd was brought over to help the Inuvialuit people because there were no more caribou here. However, the reindeer were not used for what they were intended, which was to help the people. Our elders always told us that one day the caribou would come back. In the meantime, the delta and the coastal area were full of all the other animals we harvested, from whales to birds to fish to moose. For decades, there were no or few caribou seen in the area. Eventually, they started coming back.

In the early 1970s, with six other people from Tuktoyaktuk, I went out to hunt caribou around Rufus Lake, about 130 kilometres east of Tuktoyaktuk. There were no caribou there, so we went to Anderson River. We stayed out for about seven days. Three of the people with us decided to return home without caribou. Four of us stayed for a few more days. We moved toward Mason River, about 195 kilometres east of Tuktoyaktuk, to look for caribou.

We found a herd of about fifteen caribou. We managed to get three caribou between the four of us. The caribou were very wild at the time. They were skittish around people, and we could not get too close without them running away. That was the beginning of the caribou returning. Slowly, the caribou migrated back, and since the 1970s there have been caribou around Tuktoyaktuk. We have been fortunate to have them back. A whole new generation of Inuvialuit has grown up with caribou in this area and knows only of a life with caribou. But the elders have told us that one day the caribou will leave again.

Some elders have seen this cycle happen more than once. Mrs. Gruben recalls a time it happened before. In the 1930s and 1940s, when she was young, her family stayed in a place called Nalluk, whose name means "a place where the caribou cross." She saw the caribou leave that area, and she saw them come back. Now, she is seeing them leave once again.

Most of us who grew up around Tuktoyaktuk have seen the caribou come back into our area. Also, we have seen them leave our area. Our youth do not know of a life without caribou, and they will have to learn to adjust. Our elders always told us that the caribou would once again leave and then come back. It is part of the natural cycle for the caribou. Although we need to be careful not to harm the caribou herds for future generations, our traditional Inuvialuit culture has shown us ways to sustainably harvest and manage the caribou throughout their cycle. We are also lucky to live in an area where there are many other resources to harvest, such as moose and waterfowl, both in the spring and in the fall. Also, there are lots of different types of fish. We also harvest ring and bearded seals and whales. Now that the caribou herds are again moving away, some of us will not be around to see them come back to the area, but our grandchildren will see their return, and they will know firsthand about the caribou's natural cycle.

2

The Past Facing Forward: History and Caribou Management in Northern Canada

John Sandlos

The past is always with us. On a personal and community level, history gives us common reference points that shape our present-day identity. Although subject to the distortions of nationalists and ideologues, major historical events – origins stories, conflict, revolutions, discovery narratives – reach out to us across the gulf of time to provide a sense of who we are. At the same time, history is often evoked as a guide to current policy debates, the cliché that we ought to learn from our mistakes suggesting that the past is a roadmap, however imperfect, to challenges that face us in the present (Black, 2005; Cannadine, 2008; MacMillan, 2008).

Such attempts to bridge the past and the present are fraught with danger. A second well-worn cliché – that history never repeats itself – rightly suggests that the social, economic, and ecological ground that human communities inhabit is constantly shifting. In terms of environmental policy, William Cronon has argued that history cannot provide a clear roadmap because both nature and human society are always changing. The best that environmental history can do, Cronon suggests, is provide cautionary parables about humans and nature that can inform current policy debates. One might draw lessons for the current climate change dilemma from previous efforts to mitigate air pollution problems, ranging from urban coal smoke to acid rain, but none of these local or regional issues offers a precise guide to the pervasive and unprecedented nature of the current global greenhouse gas problem (Cronon, 1993).

In more general terms, David Lowenthal has pointed to the difficulty of knowing the past, suggesting that the historical knowledge we may harness as a guide to the present day may be clouded by faulty memory, the fragmentary nature of written sources, the biases inherent to viewing history through a contemporary lens, and the fact that we can never experience the past as it truly was. The past is indeed, as Lowenthal has so famously declared, a foreign country, a repertoire of human experience that we may draw upon to create meaning in our own lives but that we can never reconstruct in its totality (Lowenthal, 1985).

However imperfect our attempts to recreate the past, historical studies can provide a powerful critique or corrective to contemporary environmental policies or conservation programs. Not only does the benefit of hindsight allow historians to reflect on the obvious shortcomings of early conservation and environmental activism, but historians are also well equipped to critique the methods that naturalists and ecologists have used to construct baseline ecological conditions, including historical wildlife abundance and vegetation cover. Perhaps the most emblematic example of history's potential as a corrective to the assumptions governing contemporary policy is the work of Melissa Leach and James Fairhead. In several seminal papers, these two social anthropologists have argued that the tendency of colonial authorities and environmental organizations to produce alarming reports of human-induced deforestation and desertification in West Africa neglected the fact that it was local people who had planted and cultivated the existing patches of forest, creating rather than destroying the sparsely wooded parklands that have come to characterize the region (Fairhead & Leach, 1996, 1998; Leach & Fairhead, 2000a, 2000b). The historian Stephen Pyne (1982, 1984, 2007) has similarly challenged contemporary paradigms of forest fire management and preservationist approaches to the management of protected areas with his prolific work on the historical place of fire in North American and global landscapes. More recently, the media called upon historical experts on natural disasters and the local environment of New Orleans, such as Craig Colten (2006) and Ted Steinberg (2005), to comment on the policy failures that contributed to Hurricane Katrina.

How, then, might history inform the policy responses to the current crisis over barren-ground caribou that emerged in Canada's Northwest Territories (NWT) and Nunavut in the mid-2000s? Are there lessons to be learned from previous caribou crises in the regions, particularly the federal government's intensive response to the spectre of mass caribou declines in the 1920s and 1930s, as well as again in the 1950s and 1960s? More precisely, how does the history of conflict between Native hunters and state wildlife managers inform current debates over caribou conservation in the NWT? The answers to these questions are not simple. Just as the past is foreign territory, so too is the complex process of present-day caribou conservation difficult to capture in its totality, particularly for a historian who is not directly involved in the process. Nonetheless, too little attention has been paid to the ways that past conflicts over wildlife management in the NWT colour current policy debates, particularly in a crisis atmosphere. If historical knowledge of previous caribou crises cannot provide a clear path out of the current challenges facing caribou conservation, as Cronon's (1993) work suggests, neither does wilful amnesia offer a constructive way forward. Certainly, we should not allow past conflict to discourage an active response to the current caribou crisis, but context-ualizing current wildlife conservation practices and conflicts in historical terms can contribute to the current policy process in two important ways. First, attentiveness to the past suggests that current wildlife conservation practices, still based largely on the authority of state managers and science, do not represent a natural hierarchy but are themselves conditioned by and produced through particular historical circumstances. Second, cur-rent conflicts between Native hunters and wildlife managers are in many ways an extension of the neocolonial relationships that have dominated the practice of state wildlife conservation in the NWT since the early twentieth century. For contemporary players involved in the process of making caribou policy, attentiveness to this history provides a means to acknowledge the political context, rather than just the managerial impera-tives, in which debates over conservation and access to northern wildlife take place. By recognizing that caribou conservation in the NWT is part of a much broader colonial history, wildlife managers and Native harvesters may also begin to imagine the means by which wildlife conservation might be more effectively decolonized, granting Native harvesters more sovereign control over the animal populations on which they depend.

Caribou Conservation: The Historical Context

It is tempting to recount the history of caribou conservation in the early twentieth century as a heroic narrative, a story of enlightened naturalists and bureaucrats acting decisively to save the caribou. With their constant invocations of the catastrophic near-extermination of the plains bison in the late nineteenth century, one could argue that early Canadian conserva-tionists had prudently incorporated the lessons of their own recent past, painfully aware that even spectacularly large populations of herd animals were not an inexhaustible resource. In many ways, the actions of the early caribou conservationists were benevolent and public-spirited, part of an effort to put the interests of the nonhuman world outside the messy affairs of human politics (Burnett, 2003; Foster, 1998).

But the early conservation movement in North America was never far removed from politics. By the early twentieth century, state, provincial, and federal governments in Canada and the United States had begun to assert legislative and managerial control over wildlife, public lands, and waterways, all in an effort to promote the conservation of fish, animals, and forests. Throughout the late nineteenth and twentieth centuries, Canadian and American governments also established forest reserves, recreational parks, wildlife refuges, and regulations to control the excessive exploitation of fish and wildlife (Bogue, 2000; Dorsey, 1998; Gillis & Roach, 1986; Loo, 2001, 2006; Nash, 1982; Sutter, 2002). In general, state conservation officials failed to consult local people, who derived at least part of their livelihoods from their local environment. In recent years, historians have written extensively on the resulting conflicts between state conservationists and local resource users in North America, focusing on themes ranging from the suppression of Native burning practices to the outright expulsion of Natives and non-Natives from national, state, and provincial parks (Binnema & Niemi, 2006; Burnham, 2000; Keller & Turek, 1998; Manore, 2007; Sandlos, 2005, 2008; Spence, 1999).

At the heart of the discord was the utilitarian idea, popularized by prominent conservationists such as US chief forester Gifford Pinchot, that the primary purpose of conservation was to manage resources for maximum productivity and the broader public good (Hays, 1959). Although state conservationists frequently levelled their critique at large-scale capitalists who mined timber or overgrazed rangelands, they also derided local subsistence users for their supposedly irrational use of natural resources. Conservationists focused not only on issues of overharvesting but also on hunting and fishing methods such as netting fish or jacklighting deer that violated the spirit of a prevailing sportsman's code steeped in the idea of the manly chase (Reiger, 1975). In broad terms, conservationists constantly depicted local subsistence users as merely following their own parochial interests; only the rational hand of state managers and scientists could see the bigger picture and harness natural resources to maximize productivity and serve the material needs of the larger society (Jacoby, 2001; Parenteau, 2004).

In northern Canada, the ideals of the sportsman's code and the state-management imperative converged around the issue of caribou conservation during what might be termed the first caribou crisis of the 1910s and 1920s. As early as the 1890s, British and American sport hunters produced sensational accounts of massive and wasteful caribou hunts carried out by northern Native hunters. In the early years of the twentieth century, an increasing number of geologists, naturalists, non-Native trappers, and mounted police officers echoed the earlier reports of the sport hunters, which decried the tendency of Native people to slaughter large numbers of caribou, particularly at river crossings, where observers held particular disdain for the supposedly gruesome and wasteful spectacle of hunters spearing highly concentrated herds (Campbell, 2004; Sandlos, 2007). In 1915 Henry Bury, a timber inspector for the Department of Indian Affairs, reported localized "extinctions" of caribou due to the improvident slaughters that Native hunters inflicted on them (Bury, 1915).

The federal government began to act on these reports in the 1910s, particularly as the Dominion Parks Branch, under the very dynamic leadership of James Harkin, gained authority over wildlife management in the Northwest Territories in 1917. The federal government's interdepartmental Advisory Board on Wildlife Protection, which included Harkin as a member, had also repeatedly identified the conservation of barrenground caribou as a priority after its creation in 1916 (Hewitt, 1918, pp. 8–9). The result was revisions to the Northwest Game Act in 1917 that created unprecedented closed seasons on caribou for Native hunters, who had previously been exempt. Although hunters could take caribou during the closed seasons if they were starving (how to distinguish starvation from hunger is not specified in the legislation), hunting barren-ground caribou was officially banned in the NWT each year in October and November and from April to the end of July (Government of Canada, 1917). In 1929 the federal government revised the game regulations to impose a ban on killing caribou cows with calves at any time of the year (Government of Canada, 1929). The enforcement of the regulations was sporadic due in part to the very small number of police in the NWT but also because government officials feared a massive relief bill if Native hunters were cut off from their most important food supply. The police and the courts successfully prosecuted only twenty-five hunters under the Northwest Game Act between 1919 and 1924, although records of these convictions are scattered and may be incomplete. Regardless, for Native hunters in the territories, the expansion of state control over local wildlife was a source of intense apprehension, prompting complaints and treaty boycotts over the closed seasons from the 1920s to the 1940s. At least on paper (and at times in practice), the closed season represented some of the strictest caribou-hunting regulations ever imposed on Native hunters in the NWT, a much reviled imposition of state control over seasonal hunting patterns that was ironically repealed in 1955 just as the much more high-profile post–Second World War caribou crisis was beginning (Sandlos, 2007).

Some of the federal government's caribou conservation policies in the 1920s and 1930s were aimed at maintaining the supply of wildlife *for* Native hunters rather than limiting access. These policies included wolf bounties, the creation of game preserves that excluded non-Native hunters, and efforts to remove trading posts along the Arctic Coast due to fears that coal smoke was interrupting the caribou migration (Usher, 2004). Although missionaries and government officials often used paternalistic language to justify such policies, imposing a kind of enforced primitivism through their desire to protect Dene and Inuit hunters and caribou from outside influences, in pragmatic terms the influx of outsiders did present a potential threat to wildlife populations that served as a base for the Native economy in the NWT. Government officials realized that, in the short term at least, completely severing Dene and Inuit hunters from their intimate connection with, and dependence on, caribou was simply not realistic. Indeed, after some internal debate in the early 1930s, the northern administration decided to maintain an important aspect of the Native cash economy, allowing Native hunters to sell caribou to non-Native outsiders who were ineligible for the general hunting licence, including the large influx of workers in the new mining camps at Port Radium and later Yellowknife (Sandlos, 2007).

At the same time, the 1910s and 1920s marked the government's first attempts to gradually shift the subsistence patterns of northern people away from the caribou. Government agents living in the Northwest Territories, such as police, Indian Agents, and game wardens, began to promote a somewhat ad hoc program of conservation education during this period. Through informal conversations and organized promotional campaigns involving postering or formal teaching within individual communities, the government promoted good-hunting practices such as avoiding the crippling loss of large numbers of caribou and not hunting bucks. Government officials also attempted to convince Native people to shift their efforts toward alternative food sources such as fish or, where feasible, moose and marine mammals (Sandlos, 2007). In 1924 the federal government's Northwest Territories and Yukon Branch hired W.H.B. Hoare, an Anglican missionary with extensive experience travelling in the Arctic, to spread conservation ideas to the Inuit who lived along the Arctic Coast. Over the next two years, Hoare covered thousands of kilometres and preached his conservation gospel to dozens of hunters, trying to convince them to give up their inland slaughter of the caribou and remain on the coast each spring and summer to fish the lakes and rivers or hunt seals and whales. Hoare (1927) claimed that his teachings had produced some tangible results in the Coronation Gulf region, where large numbers of hunters adopted salmon fishing in the late spring and early summer when they would have normally been moving inland toward the caribou herds. Whatever the veracity of Hoare's reports, clearly the government had adopted the paternalistic idea of shifting Inuit material culture from hunting to fishing, asserting control over Native subsistence patterns as a means to conserve the caribou.

In addition to fishing, the government began to formulate more radical proposals to conserve the caribou through the transformation of northern hunting cultures. After the First World War, a convergence of interests, including the federal government's northern administration, promoters such as Vilhjalmur Stefansson, and the private sector, began to imagine a future for northern Canada that included vast herds of domesticated caribou and muskox, as well as imported Old World reindeer. Inspired by the success of missionary Sheldon Jackson's experiment with reindeer introductions and Inuit herding schemes in Alaska beginning in 1892, Canadian northern boosters promoted an image of the northern tundra teeming with domesticated animals propagated through rational herd-management schemes. Native northerners, in turn, would have the opportunity to give up their supposedly random and precarious existence as hunters and adopt the lifestyle of the Christian pastoralist. This first step toward civilization would lead to others, as Arctic ranches produced the food base necessary to establish large-scale industrial activity in the region. This notion that northern progress could be built on a new ranching economy entranced federal officials to such an extent that the government attempted one unsuccessful introduction at Fort Smith, NWT, in 1911, established a royal commission to study the matter in 1919, and issued three Arctic grazing leases to private companies between 1918 and 1928, although none of these leases resulted in ranches (Diubaldo, 1978; Hewitt, 1921, pp. 136–42; Porsild, 1936; Treude, 1968). Although these were small steps, limited in success by the vagaries of financing and Arctic environmental conditions, they were the product of the thoroughly colonial idea that the introduction of domesticated reindeer and/ or the domestication of the caribou herds might spur the evolutionary advancement of the people of the Inuit and Dene hunting cultures to a more settled and civilized state as agriculturalists.

Although federal officials applied their concerns about caribou populations broadly across the NWT, the western Arctic – particularly the lower reaches of the Mackenzie River Valley – was one of the most important areas of concern and test sites of conservation activity during the first caribou crisis. Federal wildlife officials repeatedly invoked the severe depletion of the western Arctic herds (today's Porcupine caribou herd) to provision the region's whaling trade of the late nineteenth century as evidence of both Native rapaciousness and the fact that caribou herds everywhere in the NWT were threatened. The famous explorer Vilhjalmur Stefansson had in fact first raised the issue with letters to Prime Minister Robert Borden and Commission of Conservation chair Clifford Sifton in 1914, suggesting that introductions of firearms to the central Arctic would result in an assault on the caribou similar to what had occurred in the western Arctic. Obviously, Stefansson's analysis failed to account for the very specific local markets that emerged due to the large number of whalers operating out of the base at Herschel Island, but federal wildlife officials and field agents such as Hoare greatly feared that contact between American trading ships and the more isolated Inuit of the central Arctic Coast might lead to the mass slaughter of the caribou, particularly if a new market for hides and meat became available to Native hunters (Hoare, 1927; Stefansson, 1914).

The decline of the western Arctic caribou herds also made the lower Mackenzie Valley region the ideal location, in the eyes of the northern administration, for the introduction of domestic reindeer. In 1929 the federal government purchased 3,000 reindeer from an Alaskan reindeer company and introduced 2,382 of the surviving animals to a preserve in the Mackenzie Delta after a harrowing herding odyssey that lasted until 1935. The animals were meant as a replacement for what was thought to be a permanently depleted caribou herd in the region. The northern administration also hoped that the reindeer would provide stable employment for the Inuvialuit, hiring Sami herders to instruct the Inuit in their craft and encouraging the development of Native-owned herds. The project was moderately successful, never expanding beyond a small-scale cottage industry but still providing local employment for small numbers of Inuvialuit and producing inexpensive meat and hides for the local population and for distribution throughout the NWT. If such tentative results proved disappointing, the Mackenzie Delta reindeer project was originally intended as a proving ground for the thesis that irrational and wasteful hunting economies could be replaced by a more ordered and managed approach to animal husbandry (Piper & Sandlos, 2007; Treude, 1979).

Within the western Arctic, the settlement of Aklavik was one of the few in the NWT subject to relatively strict enforcement of the game regulations and vigorous promotion of conservation education. In the late 1920s, the town's chief medical officer of health, Dr. J.A. Urquhart, proved to be a zealous promoter of caribou conservation. He required Native hunters to request his direct permission to hunt caribou out of season in the event of a local food shortage. With the enthusiastic support of the northern administration, Urquhart advocated strict enforcement of the summer closed seasons in the NWT. He constantly reminded Native hunters in the Aklavik area that they should subsist on fish rather than on caribou during the summer months. Always a proponent of thrift, Urquhart complained to his superiors that local Native people were too lazy to put up an adequate supply of fish during the summer months to carry them through the fall and winter season. At least one Indian Agent, W.R.M. Truesdell at Arctic Red River, criticized Urquhart for his rigid approach, arguing that his own policy of allowing each hunter in his district to hunt two caribou during the summer months was a more realistic application of the law. Even if Urquhart's ability to supervise every hunter in the Aklavik area was limited, his actions suggest the extent to which the caribou crisis of the 1910s and 1920s provoked a determined response from the federal government (Sandlos, 2007, pp. 180–86).

The contraction of the northern administration during the Great Depression and the Second World War prevented any major advances or innovations in caribou conservation during this period. With the postwar growth in the civil service, however, caribou conservation once again became a major priority for the northern administration. A series of new "high-tech" aerial surveys conducted by the Canadian Wildlife Service beginning in 1948 produced population counts in the hundreds of thousands rather than the millions that had been assumed as a baseline population for the species. Although these surveys were a coarse scientific tool, with wide transect lines and problems with observer error preventing anything more than an educated guess in terms of caribou populations, the northern administration declared a new caribou crisis that lasted through the 1950s and 1960s (Cameron, Whitten, Smith, & Reed, 1985; Ruttan, 2012; Urquhart, 1989). As mentioned earlier, however, the northern administration recognized the futility of enforcing closed seasons or bag limits across such a vast landscape. Other than attempting to restrict the sale of caribou meat, the northern administration thus abandoned the strict regulatory approach of thirty years earlier. Instead, the northern administration concentrated its efforts on nonregulatory approaches, including an intense wolf control program, conservation education, and the production of alternative country food resources with Native communities (Usher, 2004).

If these measures seem more benign than those of the previous conservation regime, the program was in fact applied with increasing vigour as the number of federal field agents multiplied in the North. Indian Agents, police, game wardens, and the Northern Service officers who worked in Inuit communities all distributed educational material on conservation, such as filmstrips, comic books, storybooks, and posters, but they also distributed fishing nets and actively organized hunts for marine mammals. By the late 1950s, low fur prices and incidences of starvation in the eastern Arctic due to localized caribou shortages had prompted the government to advocate Native industrial employment and newly open mines as a replacement for a moribund wildlife economy and a critical component of the caribou conservation program. At the extreme end of this new policy regime, some Inuit and Dene groups and individuals were relocated away from their hunting grounds in the interior regions to the west coast of Hudson Bay, in some cases for occupational training or for work in mining projects such as the Rankin Inlet Nickel Mine. Once again, caribou conservation was linked to a more general program of modernization in the North, a reiteration of the idea that the salvation of the caribou depended on the removal and the remaking of the human hunters who had preyed on them for centuries (Bussidor & Bilgen-Reinhart, 1997; Sandlos, 2007; Tester & Kulchyski, 1994, 2007).

In spite of some contrasts in policy, there was a clear colonial dimension to both caribou crises of the twentieth century. Acting on questionable scientific evidence, the federal government used the crisis mentality that emerged in the 1920s and the 1950s as justification for asserting managerial control over not only the caribou herds but also the Native people who hunted them. State control, in turn, meant much more than regulating the hunt: the federal government also attempted to push Native hunters toward alternative resources or to completely replace their supposedly backward hunting cultures with more modern means of making a living off the land, whether through pastoral herding or industrial employment. Although the implementation of these programs was variable across the Northwest Territories and Nunavut, in almost no cases were Native hunters consulted on these broad changes to their livelihood. Frequently, they resisted through formal means such as treaty boycotts and through less formal protests such as simply ignoring the game regulations. In other instances, coercion or the authority of the law made it difficult for Native hunters to maintain control over their caribou-hunting practices. For at least half a century, however, the relationship between Native hunters and the state in northern Canada can be characterized as one of mutual contempt.

The Shifting Ground of Caribou Politics

The fundamental historical disconnect between Native hunters and the state makes it tempting to construct a second heroic narrative of caribou conservation in the Northwest Territories. One could argue that the general recognition of Native treaty rights and the value of the subsistence economy that emerged during the Berger Inquiry of the 1970s, in concert with innovative moves toward co-management and the incorporation of traditional ecological knowledge into wildlife management regimes, has rendered the "bad old days" of paternalistic wildlife conservation irrelevant to contemporary discussions. No doubt much *has* changed since the early days of wildlife conservation in the Canadian North. Dene and Inuit input into caribou management issues has been ensured by the creation of several wildlife co-management boards with significant Native representation, such as the Inuvialuit's Wildlife Advisory Committee created in 1984, the Nunavut Wildlife Management Board created as part of the broader land claim in 1993, and the Beverly and Qamanirjuaq Caribou Management Board created in 1982 as a response to another perceived regional caribou crisis. Aboriginal hunters have inserted their voices into wildlife management debates through local hunters and trappers associations and their local First Nations governments. First Nations have successfully negotiated the creation of protected areas for vital caribou habitat, most notably in the case of Ivvavik National Park, created through the Inuvialuit Final Agreement in 1984, and the adjacent Vuntut National Park, created in 1993 through the Vuntut Gwitchin First Nation Final Agreement (Peepre & Dearden, 2002). Case studies of co-management boards suggest that in some cases Native hunters feel politically engaged and empowered through their participation in creating wildlife management policy and able to play a significant role in shaping conservation programs for caribou and other species (Gertsch, Dodds, Manseau, & Amagoali, 2003;

Huntington et al., 2002; Parlee, Manseau, & Lutsel K'e Dene First Nation, 2005; Spaeder & Feit, 2005).

However, for all of these changes, there can be little doubt that colonial relations governing early northern conservation programs are still embedded in current wildlife management structures. Some studies of specific co-management boards suggest that the voices of Native hunters are secondary to those of scientists and bureaucratic managers. Several of the co-management boards afford only minority representation for Native hunters, and the powers of the boards themselves are often only advisory, with authority to actually enact regulations still residing with the relevant Cabinet minister in Nunavut or the Northwest Territories (Rodon, 1998). Anne Kendrick (2003) has observed that the co-management boards rarely adopt Aboriginal forms of decision making in their deliberations. Other observers have similarly noted that wildlife co-management meetings are often dominated by bureaucratic language, technical terms, formal presentation methods (often using PowerPoint), and physical environments such as boardrooms and conference centres that can be alienating and uncomfortable for Native participants (Nadasdy 1999, 2003a). Even pragmatic issues, such as the fact that the harvesters who sit on the Beverly and Qamanirjuaq Caribou Management Board often do not have a salary, reinforce the inequity between Native hunters and the biologists and bureaucrats who sit on the board (Kendrick, 2000). The rift between managers and harvesters can be so significant that, according to historian David Neufeld (2006), some local people in the western Arctic refer to the Porcupine Caribou Management Board as the "biologists' board," a reflection of concerns that the board does not effectively incorporate the voices of harvesters into the advisory process – although this has not precluded agreement between the board and local hunters on a broad suite of issues, as Kristine Wray suggests (see Chapter 12).

Some critics have suggested that the alienation of Aboriginal participants from the co-management process is rooted in a tendency to regard traditional ecological knowledge (TEK) as merely supplementary to Western science. Wildlife managers tend to welcome TEK as long as it is legible within a scientific and bureaucratic context, providing data on population, climate trends, and migratory routes that supplement existing bodies of research. Conflicts arise, however, when Native hunters step outside of the data-driven discourse, suggesting that animals possess their own complex social worlds or that human behaviour toward wildlife should be governed by traditional spiritual values (Cruikshank, 1998; Nadasdy, 1999, 2003b; Spak, 2005). The most commonly cited flashpoint for this divergent set of worldviews is the widespread objection of many Native hunters to the use of radio collars because it is a sign of disrespect and a potential source of harm to the animals (Kendrick, 2000). Nadasdy (2003a) has also observed that Kluane First Nations members of the Ruby Range Sheep Steering Committee objected to scientific recommendations for hunters to take only older male mountain sheep because the members saw these sheep as the elders of the herd, who were responsible for teaching younger individuals about the optimal areas to graze on a seasonal basis. Usher (2000) may be correct to note that, in pragmatic terms, efforts to incorporate TEK into formal decision-making structures should focus on its affinity with scientific knowledge, ensuring that TEK is presented in a way that is structured, verifiable, and open to criticism. Nonetheless, as many critics have noted, demands that TEK conform to the standards of Western science also run the risk of subsuming or co-opting Aboriginal knowledge within the state-management apparatus, reducing all wildlife debates to instrumental questions while ignoring the conflict of worldviews and the history of the inequitable power relations that have characterized northern wildlife conservation from the very beginning. Co-management and TEK can be understood as tools to convince Native hunters to buy in to state-management approaches, while at the same time hiding from the historical conflicts that led to these new consensus-building approaches in the first place (Ellis, 2005; Kofinas, 2005; Nadasdy, 2005).

Obviously, there is truth to the argument that co-management and the use of TEK are at least in part a corrective to the colonial past. Certainly, contemporary wildlife managers in the Canadian North no longer completely ignore Aboriginal voices; nor do they actively try to shape and control Dene and Inuit material cultures through the medium of wildlife conservation. In most cases, however, the state has not surrendered its near-monopoly of power over wildlife conservation, incorporating harvesters for the most part as participatory advisers rather than as primary decision makers within the wildlife policy process. The fundamental authority of state managers over local harvesters, one of the major principles of conservation since the rise of Pinchot's utilitarian approach in the early twentieth century, remains largely intact in northern Canada. If a complete return of Native autonomy over wildlife decisions seems unlikely due to the entrenchment of the current system, the history of colonial dispossession surrounding wildlife conservation in the North serves as a powerful reminder that contemporary management debates involve critical issues of political power, authority, and control.

Although the adherents of scientific management prefer to remove conservation issues from the realm of politics, the colonial approaches of the past undoubtedly influence current conflicts over wildlife conservation, a source of discord that is unlikely to fade unless addressed directly rather than discounted through limited consensus-building processes.

The development of a new caribou crisis in the Far North over the past several years suggests that political conflicts over managing wildlife quickly boil to the surface during times of scarcity. The precipitous drop in the population of the Bathurst caribou herd from 120,000 animals in 2006 to approximately 32,000 in 2009 prompted a strong response from the Government of the Northwest Territories. In January 2010, Environment Minister J. Michael Miltenberger imposed a caribou-hunting ban across the entire range of the Bathurst caribou herd. The Yellowknives Dene First Nation subsequently took the government to the territorial Supreme Court, challenging its authority to legislate First Nations hunting in the region. Minister Miltenberger responded in the press by suggesting that "people have been dragging out these hoary old arguments since the sixties, questioning the credibility of the government. Luckily, in this case, I think the majority of people accept the fact that the herd is in serious trouble. And when conservation hits an emergency state, we have authority" (quoted in White, 2010). At public hearings on the issue, Native and non-Native caribou users expressed legitimate concerns about the accuracy of the herd counts and about the impact of mining and other developments on caribou relative to the hunting harvest. The assistant deputy minister of the environment argued that the government had no choice but to concentrate on harvest rates because doing so would have the most immediate impact on herd numbers (Livingstone, 2010). In other words, given the economic importance of industrial development in the North, the government chose to concentrate on harvest rates as the factor that could be most readily submitted to management and control. In any case, by the fall of 2010, both the Yellowknives and the nearby Tłı̨chǫ Nation, working through the Wek'èezhìi Renewable Resources Board, had reached an agreement with the territorial government that did not impose a strict quota but instead allowed for a total First Nations target harvest of 300 caribou on the Bathurst range and 2,800 animals from the Bluenose East herd (CBC News, 2010a). Despite these concessions, for almost a year the territorial government unilaterally imposed the strictest caribou regulations ever implemented in the Northwest Territories.

Despite such extreme measures, it would be a mistake to suggest that the current caribou situation is simply a repetition of the crisis fifty years ago. The NWT government's caribou management plan for 2006–10 identifies engagement of all partners, including First Nations, as its first of five key guiding principles. In response to the dramatic decline in the barren-ground caribou herds, territorial and First Nations governments organized a multi-stakeholder caribou summit held in Inuvik in 2007, which produced a long list of recommendations for the Government of the Northwest Territories (Tesar, 2007). In the spring of 2010, eight participating governments of the Porcupine Caribou Management Board – including the Inuvialuit Game Council, the Gwich'in Tribal Council, the Vuntut Gwitchin government, the Tr'ondëk Hwëch'in First Nation, the Na-Cho Nyak Dun First Nation, the Yukon government, the NWT government, and the federal government – remarkably came to an agreement on a caribou management plan that stipulated only voluntary restrictions on a bull hunt for Aboriginal harvesters, although a harvest ban would go into effect if herd numbers dipped below 45,000 (CBC News, 2010b).

If there is a historical lesson to be drawn from the older caribou crises and applied to the current one, it is that northern Native hunters recognize the colonial legacy associated with state wildlife conservation. Throughout the past century, Dene and Inuit hunters have tended to reject wildlife conservation initiatives that solidify the power of the state over northern wildlife. Whether through treaty boycotts, lawbreaking, or more formal court challenges, northern Native hunters are likely to resist top-down wildlife conservation initiatives that undermine historical ties with and sovereignty over local animal populations. Indeed, strict regulatory approaches are as likely to provoke hunters as they are to encourage conservation, as when the *Globe and Mail* reported that the 2010 caribou ban had prompted at least one hunter to suggest he would head out to kill caribou in protest (White, 2010). The recent Porcupine caribou and Bathurst caribou agreements have stipulated voluntary conservation measures for Native hunters or for harvest targets as opposed to firm quotas, an implicit recognition of Native harvest rights that still achieves critical conservation objectives (CBC News, 2010b). As the Government of the Northwest Territories discovered, it is difficult to step outside of history and impose unilateral state control over wildlife because First Nations that are pressing for greater local authority and self-government simply do not accept the imposition of external control over wildlife populations they have hunted for generations. The symbolism attached to the language of wildlife policy is critical: a target harvest, for instance, as opposed to a quota, affirms Aboriginal sovereignty over wildlife even if the material outcome – a very low caribou harvest of 300 – is the same in either case. A report by the Wek'èezhìi Renewable Resources Board (2010) on the Bathurst caribou herd emphasizes the Tłįchǫ Nation's desire to make responsible decisions about the caribou within a context of personal and community autonomy. In a crisis situation, it may be tempting for state wildlife authorities to fall back on familiar authoritarian patterns that provide seemingly simple solutions to complex problems. It is moments of crisis, however, that provide the most meaningful test of the dominant society's willingness to incorporate Native input into and devolve authority over caribou management processes.

As much as history may help us to understand the political context for northern wildlife conservation, it cannot provide any guarantee that current conservation efforts will be successful. Evidence from science and TEK does suggest that caribou populations cycle in oscillations of roughly thirty to forty years, a history of declines and irruptions that may alleviate concern about the current crisis if they can be contextualized as something natural (Beaulieu, 2012; Ferguson & Messier, 1997; Ferguson, Williamson, & Messier, 1998). Nonetheless, many observers of the current crisis have implicitly invoked Cronon's (1993) arguments about the unreliability of history as a guide to the future, suggesting that industrial development and climate change are unprecedented influences on the caribou that may lead to their permanent diminishment or demise (Gunn, Russell, White, & Kofinas, 2009). Mining industry representatives have countered that current mining projects have had no adverse impacts on caribou (Northwest Territories Chamber of Mines, 2010). However, the proliferation of ice roads on the tundra may interrupt migratory patterns, and they certainly provide hunters easier access to the caribou herds. One scientific study suggests that caribou tend to avoid industrial development areas within a five-kilometre radius (Vistnes & Nellemann, 2008). Combine these impacts with a changing climate that could produce very rapid alterations in vegetation patterns on tundra summer ranges, an increase in forest fires in the boreal winter range that would reduce lichen cover, or an increase in the number of insects harassing the herds in summer, and the future of the barren-ground caribou herds could be very bleak (Sharma, Couturier, & Côté, 2009; Vors & Boyce, 2009).

Ultimately, however, after fifty years of scientific research and more recent studies using TEK, we still do not know for certain why caribou herds are subject to large-scale periodic declines. Several localized studies suggest that weather events such as large amounts of snow or ice cover

can cause declines in caribou populations because they make it difficult
for the animals to access lichen (Miller & Barry, 2009; Miller & Gunn,
2003). Conversely, one study has questioned the idea that weather is a
primary limiting factor, suggesting that large amounts of snow cover and
warming temperatures can produce increases in local caribou populations (Tyler, 2010). The large gaps in our knowledge of caribou population dynamics reinforce the notion that neither science nor TEK has ever
provided an absolutely clear roadmap that can be used to guide human
management of the caribou herds. As much as the language surrounding
power sharing and co-managing caribou is critical to a more just application
of human politics, it should never mask the fact that the herds may
not be manageable in a conventional sense. There is simply too much
we don't know about the seasonal movements and population ecology
of caribou.

In a broader context, state management of nature through the application of scientific knowledge should never be accepted as a natural arrangement, as it is the product of a very particular and very early moment in
the political history of the conservation movement in North America.
Moreover, historical reflection suggests that the human effort to manage nature scientifically has produced its share of failed policies, perhaps
most notably the misguided policies on forest fire suppression and preda-
tor eradication of the early twentieth century. Bavington's (2010) analysis of the cod fishery's collapse in Newfoundland argues that reductive
forms of managerial ecology based on one-dimensional scientific models
of ecosystem dynamics were a key structural cause of one of the great
ecological catastrophes of recent history. Invoking the literature on resilience and adaptive management, Bavington advocates new approaches to
natural resource management that embrace complexity, uncertainty, and
a willingness to learn constantly and humbly as environmental conditions change – a broad approach to conservation that finds sympathetic
echoes in many of the chapters in the current volume. As much as state
management may still be the dominant approach in many provincial, territorial, state, and federal resource bureaucracies in North America, it has
never provided an effective approach to caribou management in northern
Canada. For the conceivable future, Native resistance to colonial wildlife
policies and the elusive and often incomprehensible nature of the barren-
ground caribou's ecology will always undermine efforts to manage the
species. This may not be a satisfying conclusion for those who would
find comfort in the familiar logic of governmental command and control. History teaches us, however, that the conservation of barren-ground

caribou must always involve a degree of muddling through, an attempt to preserve the herds while recognizing that both our knowledge base and our attempts to escape the colonial politics of the past are incomplete.

REFERENCES

Bavington, D. (2010). From hunting fish to managing populations: Fisheries science and the destruction of Newfoundland cod fisheries. *Science as Culture, 19*(4), 509–528.

Beaulieu, D. (2012). Dene traditional knowledge about caribou cycles in the Northwest Territories. *Rangifer, 32*(2), 59–67.

Binnema, T., & Niemi, M. (2006). "Let the lines be drawn now": Wilderness, conservation and the exclusion of Aboriginal people from Banff National Park in Canada. *Environmental History, 11*(4), 724–750.

Black, J. (2005). *Using History*. London: Hodder Education.

Bogue, M. (2000). *Fishing the Great Lakes: An Environmental History, 1783–1933*. Madison: University of Wisconsin Press.

Burnett, J.A. (2003). *A Passion for Wildlife: The History of the Canadian Wildlife Service*. Vancouver: UBC Press.

Burnham, P. (2000). *Indian Country, God's Country: Native Americans and the National Parks*. Washington, DC: Island.

Bury, H. (1915). *Report on the Game and Fisheries of Northern Alberta and the Northwest Territories*. RG 85, vol. 664, file 3910, part 2, Library and Archives Canada.

Bussidor, I., & Bilgen-Reinhart, U. (1997). *Night Spirits: The Story of the Relocation of the Sayisi Dene*. Winnipeg: University of Manitoba Press.

Cameron, R.D., Whitten, K.R., Smith, W.T., & Reed, D.J. (1985). Sampling errors associated with aerial transect surveys of caribou. In T. Meredith & A. Martell (Eds.), *Proceedings of the Second North American Caribou Workshop* (pp. 273–83). Montreal: Centre for Northern Studies and Research, McGill University.

Campbell, C. (2004). A genealogy of the concept of 'wanton slaughter' in Canadian wildlife biology. In D.G. Anderson & M. Nutall (Eds.), *Cultivating Arctic Landscapes: Knowing and Managing Animals in the Circumpolar North* (pp. 154–171). New York: Berghahn Books.

Cannadine, D. (2008). *Making History Now and Then: Discoveries, Controversies and Explorations*. New York: Palgrave Macmillan.

CBC News (2010a). Bathurst caribou plan to help preserve herd. October 11. http://www.cbc.ca/news/canada/north/bathurst-caribou-plan-to-help-preserve-herd-1.886027.

CBC News (2010b). Porcupine caribou management plan awaits signatures. May 6. http://www.cbc.ca/news/canada/north/porcupine-caribou-management-plan-awaits-signatures-1.910142.

Colten, C.E. (2006). Conspiracy of the levees: The latest battle of New Orleans. *World Watch, 19*(5), 8–12.

Cronon, W. (1993). The uses of environmental history. *Environmental History Review, 17*(3), 1–22.

Cruikshank, J. (1998). *The Social Life of Stories: Narrative and Knowledge in the Yukon Territory*. Lincoln: University of Nebraska Press.

Diubaldo, R. (1978). *Stefansson and the Canadian Arctic*. Montreal, Kingston: McGill-Queen's University Press.

Dorsey, K. (1998). *The Dawn of Conservation Diplomacy: U.S.-Canadian Wildlife Protection Treaties in the Progressive Era*. Seattle: University of Washington Press.

Ellis, S.C. (2005). Meaningful consideration? A review of traditional knowledge in environmental decision making. *Arctic, 58*(1), 66–77.

Fairhead, J., & Leach, M. (1996). *Misreading the African Landscape: Society and Ecology in a Forest-Savanna Mosaic*. Cambridge, UK: Cambridge University Press.

Fairhead, J., & Leach, M. (1998). *Reframing Deforestation: Global Analyses and Local Realities with Studies in West Africa*. New York: Routledge.

Ferguson, M.A.D., & Messier, F. (1997). Collection and analysis of traditional ecological knowledge about a population of Arctic tundra caribou. *Arctic, 50*(1), 17–28.

Ferguson, M.A.D., Williamson, R.G., & Messier, F. (1998). Inuit knowledge of long-term changes in a population of Arctic tundra caribou. *Arctic, 51*(3), 201–219.

Foster, J. (1998). *Working for Wildlife: The Beginning of Preservation in Canada* (2nd ed.). Toronto: Univeristy of Toronto Press.

Gertsch, F., Dodds, G., Manseau, M., & Amagoali, J. (2003). Recent experiences in cooperative management and planning for Canada's northernmost national park: Quttinirpaaq National Park on Ellesmere Island. Paper presented at Making Ecosystem-Based Management Work: 5th International Conference on Science and Management of Protected Areas, Victoria.

Gillis, P., & Roach, T. (1986). The American influence on conservation in Canada. *Journal of Forest History, 30*(4), 160–174.

Government of Canada (1917). An Act Respecting Game in the Northwest Territories of Canada, 7–8 George V, vol. 1, c. 36, s. 1. *Statutes of Canada,* 337–43.

Government of Canada (1929). Order in Council P.C. 807. *Canada Gazette, 67*(47), 2079.

Gunn, A., Russell, D., White, R., & Kofinas, G. (2009). Facing a future of change: Wild migratory caribou and reindeer. *Arctic, 62*(3), iii–iv.

Hays, S. (1959). *Conservation and the Gospel of Efficiency: The Progressive Conservation Movement, 1890–1920*. New York: Antheneum.

Hewitt, C.G. (1918). *Conservation of Wild Life in Canada in 1917: A Review*. Ottawa: Canadian Commission of Conservation.

Hewitt, C.G. (1921). *The Conservation of the Wild Life of Canada*. New York: Charles Scribner's Sons.

Hoare, W.H.B. (1927). *Report of Investigations Affecting Eskimo and Wild Life, District of Mackenzie, 1924–1925–1926*. Ottawa: Northwest Territories and Yukon Branch, Department of the Interior.

Huntington, H., Brown-Schwalenberg, P., Frost, K., Fernandez-Gimenez, M., Norton, D., & Rosenberg, D. (2002). Observations on the workshop as a means of improving communication between holders of traditional and scientific knowledge. *Environmental Management, 30*(6), 778–792.

Jacoby, K. (2001). *Crimes against Nature: Squatters, Poachers, Thieves, and the Hidden History of American Conservation*. Berkeley: Unviersity of California Press.

Keller, R., & Turek, M. (1998). *American Indians and the National Parks*. Tucson. University of Arizona Press.

Kendrick, A. (2000). Community perceptions of the Beverly-Qamanirjuaq Caribou Management Board. *Canadian Journal of Native Studies, 20*(1), 1–33.

Kendrick, A. (2003). The flux of trust: Caribou co-management in northern Canada. *Environments, 31*(1), 43–59.

Kofinas, G. (2005). Caribou hunters and researchers at the co-management interface: Emergent dilemmas and the dynamics of legitimacy in power sharing. *Anthropologica, 47*(2), 179–196.

Leach, M., & Fairhead, J. (2000a). Challenging neo-Malthusian deforestation analyses in West Africa's dynamic forest landscapes. *Population and Development Review, 26*(1), 17–43.

Leach, M., & Fairhead, J. (2000b). Fashioned forest pasts, occluded histories? International environmental analysis in West African locales. *Development and Change, 31*(1), 35–59.

Livingstone, A. (2010). NWT defends caribou ban. *Northern News Service,* March 24.

Loo, T. (2001). Making a modern wilderness: Conserving wildlife in twentieth-century Canada. *Canadian Historical Review, 82*(1), 92–121.

Loo, T. (2006). *States of Nature: Conserving Canada's Wildlife in the Twentieth Century.* Vancouver: UBC Press.

Lowenthal, D. (1985). *The Past Is a Foreign Country.* Cambridge, UK: Cambridge University Press.

MacMillan, M. (2008). *The Uses and Abuses of History.* Toronto: Penguin.

Manore, J. (2007). Contested terrains of space and place: Hunting and the landscape known as Algonquin Park, 1890–1950. In J. Manore (Ed.), *The Culture of Hunting in Canada* (pp. 121–147). Vancouver: UBC Press.

Miller, F.L., & Barry, S.J. (2009). Long-term control of Peary caribou numbers by unpredictable, exceptionally severe snow or ice conditions in a non-equlibrium grazing system. *Arctic, 62*(2), 175–89.

Miller, F.L., & Gunn, A. (2003). Catastrophic die-off of Peary caribou on the western Queen Elizabeth Islands, Canadian High Arctic. *Arctic, 56*(4), 381–390.

Nadasdy, P. (1999). The politics of TEK: Power and the "integration" of knowledge. *Arctic Anthropology, 36*(1–2), 1–18.

Nadasdy, P. (2003a). *Hunters and Bureaucrats: Power, Knowledge, and Aboriginal-State Relations in the Southwest Yukon.* Vancouver: UBC Press.

Nadasdy, P. (2003b). Reevaluating the co-management success story. *Arctic, 56*(4), 367–380.

Nadasdy, P. (2005). The anti-politics of TEK: The institutionalization of co-management discourse and practice. *Anthropologica, 47*(2), 215–232.

Nash, R. (1982). *Wilderness and the American Mind* (3rd ed.). New Haven, CT: Yale University Press.

Neufeld, D. (2006). Hunting caribou, managing caribou: Orality in the 21st century: Inuit discourse and practices. Paper presented at the 15th Inuit Studies Conference, Paris.

Northwest Territories Chamber of Mines. (2010). *Caribou Management Actions in Wek'èezhìi.* Yellowknife: Wek'èezhìi Renewable Resource Board.

Parenteau, B. (2004). A "very determined opposition to the law": Conservation, angling leases, and social conflict in the Canadian Atlantic salmon fishery, 1867–1914. *Environmental History Review, 9*(3), 436–463.

Parlee, B., Manseau, M., & Lutsel K'e Dene First Nation (2005). Understanding and communicating about ecological change: Denesoline indicators of ecosystem health. In F. Berkes, R. Huebert, H. Fast, M. Manseau, & A. Diduck (Eds.), *Breaking Ice: Integrated Ocean Management in the Canadian North* (pp. 165–82). Calgary: University of Calgary Press.

Peepre, J., & Dearden, P. (2002). The role of Aboriginal peoples. In P. Dearden & R. Rollins (Eds.), *Parks and Protected Areas in Canada: Planning and Management* (2nd ed., pp. 323–353). Toronto: Oxford University Press.

Piper, L., & Sandlos, J. (2007). A broken frontier: Ecological imperialism in the Canadian North. *Environmental History, 12*(4), 759–795.

Porsild, A.E. (1936). The reindeer industry and the Canadian Eskimo. *Geographical Journal, 88*(1), 1–17.

Pyne, S. (1982). *Fire in America: A Cultural History of Wildland and Rural Fire.* Princeton, NJ: Princeton University Press.

Pyne, S. (1984). *Introduction to Wildland Fire: Fire Management in the United States.* New York: Wiley.

Pyne, S. (2007). *Awful Splendour: A Fire History of Canada.* Vancouver: UBC Press.

Reiger, J. (1975). *American Sportsmen and the Origins of Conservation.* New York: Winchester.

Rodon, T. (1998). Co-management and self-determination in Nunavut. *Polar Geography, 22*(2), 119–135.

Ruttan, R. (2012). New caribou crisis – then and now. *Rangifer, 32*(20), 85–102.

Sandlos, J. (2005). Federal spaces, local conflicts: National parks and the exclusionary politics of the conservation movement in Ontario, 1900–1935. *Canadian Historical Association Journal, 16*(1), 293–318.

Sandlos, J. (2007). *Hunters on the Margin: Native People and Wildlife Conservation in the Northwest Territories.* Vancouver: UBC Press.

Sandlos, J. (2008). Not wanted in the boundary: The expulsion of the Keeseekoowenin Ojibway band from Riding Mountain National Park. *Canadian Historical Review, 89*(3), 189–222.

Sharma, S., Couturier, S., & Côté, S. (2009). Impacts of climate change on the seasonal distribution of migratory caribou. *Global Change Biology, 15*(10), 2549–2562.

Spaeder, J., & Feit, H.A. (2005). Co-management and Indigenous communities: Barriers and bridges to decentralized resource management. *Anthropologica, 47*(2), 147–54.

Spak, S. (2005). The position of Indigenous knowledge in Canadian co-management organizations. *Anthropologica, 47*(2), 233–246.

Spence, M. (1999). *Dispossessing the Wilderness: Indian Removal and the Making of the National Parks.* New York: Oxford University Press.

Stefansson, V. (1914). Letter to Prime Minister Robert Borden, 8 January. MG 36 H, Borden Papers 785, Library and Archives Canada.

Steinberg, T. (2005). A natural disaster, and a human tragedy. *Chronicle of Higher Education, 53*(5), B11–B12.

Sutter, P. (2002). *Driven Wild: How the Fight against Automobiles Launched the Modern Wilderness Movement.* Seattle: University of Washington Press.

Tesar, C. (2007). Canada's disappearing caribou. Arctic Indigenous People's Secretariat, February 3. No longer online.

Tester, F., & Kulchyski, P. (1994). *Tammarniit (Mistakes): Inuit Relocation in the Eastern Arctic, 1939–63.* Vancouver: UBC Press.

Tester, F., & Kulchyski, P. (2007). *Kiumajut (Talking Back): Game Management and Inuit Rights, 1900–70.* Vancouver: UBC Press.

Treude, E. (1968). The development of reindeer husbandry in Canada. *Polar Record, 14*(88), 15–19.

Treude, E. (1979). Forty years of reindeer herding in the Mackenzie Delta, NWT. *Polarforschung, 45*(2), 121–138.

Tyler, N.C. (2010). Climate, snow, ice, crashes, and declines in populations of reindeer and caribou. *Ecological Monographs, 80*(2), 197–219.

Urquhart, D. (1989). History of research. In E. Hall (Ed.), *People and Caribou in the Northwest Territories* (pp. 95–101). Yellowknife: Department of Renewable Resources, Government of the Northwest Territories.

Usher, P.J. (2000). Traditional ecological knowledge in environmental assessment and management. *Arctic, 53*(2), 183–193.

Usher, P.J. (2004). Caribou crisis or administrative crisis? Wildlife and Aboriginal policies on the barren grounds of Canada, 1947–60. In D.G. Anderson & M. Nutall (Eds.), *Cultivating Arctic Landscapes: Knowing and Managing Animals in the Circumpolar North* (pp. 172–199). New York: Berghahn Books.

Vistnes, I., & Nellemann, C. (2008). The matter of spatial and temporal scales: A review of reindeer and caribou response to human activity. *Polar Biology, 31*(4), 399–407.

Vors, L.S., & Boyce, M.S. (2009). Global declines of caribou and reindeer. *Global Change Biology, 15*(11), 2626–2633. https://doi.org/10.1111/j.1365-2486.2009.01974.x

Wek'èezhìi Renewable Resources Board (2010). *Report on a Public Hearing Held by the Wek'èezhìi Renewable Resources Board 22–26 March 2010, 5–6 August 2010, Behchokö, NT, and Reasons for Decisions Related to a Joint Proposal for the Management of the Bathurst Caribou Herd.* Behchokö, NT: Wek'èezhìi Renewable Resources Board.

White, P. (2010). NWT Natives fight to hunt dwindling caribou herd. *Globe and Mail,* May 4

3

Recounting Caribou

Brenda Parlee

B arren-ground caribou are among the most dynamic species in the circumpolar Arctic, regularly showing fluctuations in population more dramatic than any other ungulate (Bergerud, 1996). In Canada, the ups and downs of populations have been scientifically documented and critically theorized only since the late 1970s. Aerial surveys provide some useful contemporary data, but the outcomes and interpretations can be ambiguous and difficult to explain. As a consequence, caribou counts have become highly contentious in many parts of the North. Northern Aboriginal peoples, however, whose perspectives on the comings and goings of caribou are drawn from generations of accumulated traditional knowledge, may be able to add to our understanding of caribou population dynamics in ways that contribute to management.

This chapter presents narratives about changes in caribou populations in the western Canadian Arctic drawn from research with five communities in the Inuvialuit, Gwich'in, and Sahtú regions. The patterns and perspectives of these regions are linked to other sources of previously documented traditional knowledge found in other areas of the Northwest Territories, including the Tłı̨chǫ region and the lands of the Łutsël K'e Dene First Nation. To provide context for these perspectives and their significance, the chapter begins with an overview of some key scientific theories and mainstream methods for assessing caribou population change. The aim of the chapter is to suggest how traditional knowledge about caribou and caribou-human relations can improve our understanding of population dynamics in ways that are distinct from conventional caribou "counting."

Background: Causes of Changing Populations

The Porcupine, Cape Bathurst, Bluenose West, and Bluenose East herds have a combined range of over 750,000 kilometres, with migrations distancing between 700 and 2,500 kilometres annually. Given such vast geography, it is no surprise that efforts to calculate herd size or to "count caribou" have been problematic. Counting caribou has nonetheless become a preoccupation of many northern governments. The underlying model of caribou management is not unlike other wildlife management and conservation planning processes elsewhere in Canada and globally, which tend to be narrowly defined, technology dependent, "expert-driven," deterministic, and target-oriented (Riley et al., 2002). "A major limitation of these approaches is that there is no formal way of accounting for uncertainty in the inputs, or representing uncertainty about the conclusions" (Thomas et al., 2005, p. 20). Indeed, at one point, critics challenged that "wildlife science was likely to collapse under the weight of unreliable knowledge," as much management decision making seemed to rely more heavily on opinion and "guess work" than on evidence (Romesburg 1981).

As a result of the failures of conventional wildlife science, many scholars see a new paradigm emerging in wildlife management, which recognizes that biological sciences continue to be relevant but are "not a sufficient stand-alone basis for the practice of wildlife management" (Riley et al., 2002, p. 586). As a consequence, there is increasing pressure on the old guard of wildlife management experts to be more multidisciplinary, participatory, and more inclusive of other ways of knowing.

Nowhere is this tension more evident than in counting caribou. For some Aboriginal peoples who believe animals to be sacred and among their familial relations, many aspects of conventional caribou management are too narrow or limited in their approach to understanding caribou population dynamics and other aspects of caribou ecology (Legat, 2012). More specifically, the preoccupation with the arithmetic of counting caribou is antithetical to other ways of knowing that hinge on respect and reciprocity. Surveillance of caribou, according to Foucauldian scholars in environmental politics and elsewhere, is among the ways that experts and governments have exerted control or expressed power or sovereignty over a vast and uncertain Arctic environment (Huebert, 2011; Nadasdy, 2005; Rinfret, 2009). The accumulation of knowledge through highly technical and costly processes such as aerial surveys, may also be theorized as a means of subjugating Aboriginal peoples who have little access to similar technology and resources and whose traditional knowledge has historically been little recognized or valued (Nadasdy 2005).

Scientists have been studying changing caribou populations for decades, with thirty- to seventy-year population cycles being theorized as the norm for most herds in Canada and Alaska (Bergerud, Jakimchuk, & Carruthers, 1984; Kelsall, 1968). Only in recent decades, however, has detailed scientific evidence been gathered regarding possible triggers of that change, with habitat disturbance being a major consideration. Barren-ground caribou, like many other species, are limited by the carrying capacity of their range, including the availability of forage. Some researchers have suggested that over decades, caribou populations are regulated by density-dependent forage depletion (Ferguson, Gauthier, & Messier, 2001). In simple terms, if numbers exceed the availability of forage, caribou will experience declines in body condition, with population decline eventually being caused by disease (i.e., parasites), low fertility in cows, and poor calf survival over several years (Albon et al., 2002; Kutz, Elkin, Panayi, & Dubey, 2001; Messier, Huot, Le Henaff, & Luttich, 1988). Over time, the land will recover, with caribou numbers rebounding as forage and calf recruitment increase. However, repeated or chronic changes in the availability or condition of habitat are a key focus of research given compounded increases in large-scale resource development (Boulanger et al., 2012; Brotton & Wall, 1997; Cameron, Lenart, Reed, Whitten, & Smith, 1995; Crete & Payette, 1990; Gunn, Russell, White, & Kofinas, 2009; Johnson et al., 2005; Russell, van de Wetering, White, & Gerhart, 1996).

Short-lived effects can be associated with extreme weather events such as a late frost, rain on snow (e.g., freezing rain), or unseasonable heat. The North Atlantic Oscillation effect is theorized as the culprit behind synchronous declines in caribou populations and wildlife throughout the North American Arctic (Aanes et al., 2002). Such influences can lead to poor calf survival or to a decline in condition (i.e., skinny animals), as well as to increased mortality, but they are unlikely to have drastic or long-term effects. Predation by wolves and grizzly bears also influences caribou numbers; however, the predator-prey balance is considered synchronous, with the numbers of bears and wolves decreasing with any decline in caribou (Bergerud, 1996). Forest fires can have a major influence in some parts of the Northwest Territories. In northern Quebec, however, the effect of forest fires was shown to be minor and short-lived, with the influence being more on caribou migration than on caribou numbers (Payette et al., 2004).

Harvesting or "overharvesting" is presented as a major problem in many areas. When the number of animals taken per annum exceeds population recruitment (i.e., the number of calves born that survive each year), the population is anticipated to decline. The harvesting of cows is considered by biologists to be particularly problematic, as the loss of a cow or a pregnant cow amounts to reduced reproductive capacity and a compounded decline (Boulanger et al., 2011). It is this issue of harvesting, more than any other known influence on population, that has attracted the attention of policy makers and caribou managers. There are no regional or herd-wide data available about harvesting levels for caribou; existing data for the western Arctic suggest an exponentially lower harvest per annum than that calculated in the 1950s, when caribou were almost the sole source of food (Ruttan, 2012, p. 100; Sandlos, 2004). Although these harvest data from the Inuvialuit, Gwich'in, and Sahtú harvest studies have limitations, the thesis of a downward trend in harvesting is supported by public health evidence on consumption of caribou meat among Aboriginal communities over the past half-century (Kuhnlein et al., 2004). Given this body of evidence, governments seem too quick to blame harvesters and too slow to address the other causes of declines, including habitat degradation and loss due to resource development and climate change.

Caribou Counts

Counts of the Porcupine, Bluenose West, Bluenose East, and Cape Bathurst herds are by and large generated from aerial surveys that have been undertaken from time to time since the early 1970s. Aerial surveys are not a precise tool, with biologists often presenting a population range that has a declared significant margin of error. The ambiguities in conclusions have left governments and academic biologists alike open to critique and even litigation. As summed up by one biologist, "Many aerial surveys designed to estimate numbers of caribou lack clear objectives, are inaccurate and imprecise, lack application and often are doubted by the public" (Thomas, 1996, p. 15).

The methods for estimating population have varied over the past half-century. Early counts were based on the "standard strip census" method and were conducted in the fall and winter ranges. However, since the mid-1970s, best practice for estimating population has become the "aerial photo census" method, with counts being conducted in spring calving grounds. Using this method, the earliest estimate of the population of the Porcupine herd was made in 1972; numbers for the Bluenose West herd were first calculated in 1992, with the first photo census of the Bluenose East herd carried out in 2000. The aerial photo census technique varies but is generally as follows:

> All aircraft fly several thousand feet above ground level while biologists look for caribou and listen for the radio collars. A nine-by-nine aerial camera has been mounted on the belly of a De Havilland Beaver plane owned by the Alaska Department of Fish and Game. Once large groups of caribou are located by the smaller planes, this plane flies transect over the groups and takes photos at regular intervals. Smaller groups of caribou are either counted or photographed from the other search planes. The actual census usually takes one to three days. Often, waiting for the caribou to form groups takes the longest time. Large groups can form very suddenly and break up just as fast; therefore, the crew needs to be ready to go on very short notice. The photos are developed during the summer, and a number of agencies help count the caribou in the photos. (PCMB, 2011, p. 2)

Owing to improvements in technology, the science of aerial surveys has evolved over the past decades, yet significant margins of error persist, with some questioning the usefulness of the data altogether (Thomas, 1996). These margins of error come from several tenuous methodological and ecological assumptions. Most biologists and governments involved in or initiating population surveys begin with a degree of confidence that caribou numbers can be accurately determined from studying aerial photographs; however, some caribou users perceive the study of photographs as too abstract and have often raised concerns about the satellite collaring and the fly-over method being stressful for cows and young calves (Kendrick, 2002).

Population estimates are also based on assumptions that all caribou of a particular herd can be found in the calving ground at the time of the census and that caribou retain fidelity to calving grounds year after year. Beliefs about use and fidelity to calving grounds are largely grounded in radio collar evidence; but less than 1 percent of animals per herd have been collared. Furthermore, the limited availability of dollars for wildlife research means that census efforts rarely involve flying over all of the range and possible calving areas (Spak, 2005, p. 238). Despite these limitations, the results of these photo census surveys have become the most important source of information used in caribou management decision making at the local, regional, and territorial levels. The uncertainties of these assumptions, coupled with the short timeframe in which studies must be conducted, suggest there is room for more knowledge and reflexivity in decision making. In this context, the chapter examines the availability of traditional knowledge about caribou from research compiled between 2007 and 2011.

CHANGES IN THE AVAILABILITY OF CARIBOU: NORTHWEST TERRITORIES PERSPECTIVES

The relationships between northern Aboriginal peoples and caribou are social, economic, and cultural as well as deeply spiritual (see Chapter 6). The Gwich'in and Slavey peoples of the Sahtú are among those throughout the Northwest Territories who have complex beliefs about the familial relations between people and caribou. Communities from the Tłı̨chǫ and Denesǫłiné regions also have strong spiritual and cultural relationships to caribou that inform their understanding of the comings and goings of caribou. For example, the late Denesǫłiné elder Zepp Casaway of Łutsël K'e told a story called "The Caribou and the Tiny Tiny Man," which describes how a Dene man was born from a caribou hoof (Parlee, Basil, & Drybones, 2001). The "Legend of the Caribou Boy" tells of a strong physical and spiritual connection between Aboriginal people and the caribou (Blondin & Blondin, 2009). Some people describe this relationship similarly to that of a family or relatives: "My father and my grandfather, they spoke about it, and they used to tell us, like, the habits and what was going on with the caribou. They also told us that the caribou knows Deline as people, our thoughts, how we want to, like what we want to do, like, with them as animals" (quoted in SRRB, 2007, p. 117).

In the Tłı̨chǫ region, oral histories refer to a time when caribou and the Dene people spoke the same language and could understand each other. In other oral histories, caribou are referred to more as spiritual leaders who understand the future:

The ʔekwö are not human. They are not human, but like prophets they can foresee everything that's on this part of the land. They don't talk, they don't understand one another but still, that's the way they roam on the land ... As for the ʔekwö leader who they follow, she was born with the grace of God and it is like she knows what is up ahead of them. That's the way it is with the ʔekwö. In the old timer's way, they're like our relatives and we

depend on them, so we are really happy. In the same way, they know they
will not live but they are happy too. (Rosalie Drybones, quoted in Legat
et al., 2001, p. 1)

For many communities, the lifecycle of barren-ground caribou is seen as
directly interrelated and in balance with other elements of the ecosystem,
including harvesters. Some Dene hunters emphasize that this is not a predator-
prey relationship but is more spiritual in nature. Whether the hunter is
successful depends on the degree of respect shown to the animal and on
other aspects of one's relationship with the land, the community, and the
Creator. In that context, harvesting is not a threat or act of predation, as it
is portrayed in standard conservation biology texts; rather, it is the caribou
who decides to give itself as a gift to the hunter (Smith, 2002).

Furthermore, people perceive the harvest and use of the land as good
for caribou. One common view is that "the caribou come back to us
because they need us." This notion is expressed by an elder of the Sahtú
region: "The land is a living thing. If you don't use the land, it's not alive.
So the caribou knows that. He knows you have to live off it to survive out
there" (quoted in SRRB, 2007, p. 61). These beliefs about the relationship between people and caribou are fundamental to understanding local
observations about the comings and goings of caribou in any given year.
In some cases, these comings and goings reflect changes in distribution.
In other cases, they are perceived as issues of health and population.

Changes in Distribution

Few harvesters and elders are comfortable predicting where and when
caribou are likely to travel, particularly in fall and winter months. As noted
by a Sahtú elder, "The caribou have their own mind. They live the way
they want to and they travel where they want to too and they don't live
by man" (quoted in SRRB, 2007, p. 117). The reciprocal relationship
between people and caribou often factors into explanations of why the
caribou migrate to certain places year after year and why they suddenly
might avoid some areas. People depend on caribou migrating to certain
places year after year – they anticipate where and when they will see caribou
and in what kinds of numbers at well-known caribou crossings, passes,
trails, and habitats. Caribou come back to these well-known places for
many reasons. For some Dene, the caribou return because they are coming
to see the people. For example, the late Denesǫłiné elder Zepp Casaway

said that the caribou come to his area because they know the people miss them and need them. This view seems consistent with accounts from other regions among people such as the Tłįchǫ; as expressed eloquently by a Tłįchǫ elder, "It is said that, when they [the caribou] see the people for the first time, they are really, really happy" (Rosalie Drybones, quoted in Legat et al., 2001, p. 1). For some people, the movements of caribou are the result of prayer; when people pray about the caribou to the Creator, they will come: "The caribou are like the creator, when they know you need them they will come to you; when you are alone and you pray to them they will come and you will have food and clothing. Like the creator they take care of us. When they know you are in need they will help you" (Georgina Chocolate, quoted in Legat et al., 2001, p. 1). Despite this tacit acceptance of their unpredictable nature, many elders and hunters have good historical records (i.e., oral histories) of caribou distributions dating back to earlier in the century. Coupled with contemporary observation, oral histories can reveal significant detail about the changes in distribution over time. In some cases, more than 100 years of knowledge on distribution has been documented (see Chapter 1).

There are many reasons why caribou may not return to a given area. Lack of respect for caribou is one key issue that is discussed at the local level. Other interpretations offered relate to changes in the condition of the range. Among the issues of greatest and pervasive concern is the increasing presence of resource development. These concerns date back decades in many regions among people such as the Sahtú, whose first experience with oil and gas exploration was in the 1950s. During the Berger Inquiry of the 1970s, a harvester from Colville Lake shared his observations of the effect of exploration on caribou movements: "There used to be a lot of caribous. Used to go by dog team around halfway to Colville ... [and] get a lot of caribou. But since the seismic and all this choppers and all that start coming around, caribou are getting kind of scarce" (quoted in SRRB, 2007, p. 97). As described above, the caribou not coming is sometimes interpreted as an issue of distribution. In other cases, the lack of availability of caribou is considered to be an issue of poor caribou health.

CHANGES IN POPULATION: ARE THE CARIBOU HEALTHY?

Variability in both numbers and distribution is characteristic of most barren-ground caribou herds. There are many different experiences of variability and change in the different herds in different regions across the

Northwest Territories. Early recollections of caribou cycles suggest the caribou in the Kitikmeot region were increasing or peaking at the time the white men arrived (Thorpe et al., 2001). There is a general view that variability is part of a natural lifecycle and that the caribou will come back (see Chapter 1):

> The cycle of caribou, just like the human being. We die and there's another person born, that's what the elders were saying, with animals it's the same thing. Just like rabbits, they disappear one year and they come back again. It's a cycle thing, they said. And they say that's the way for us, I mean, there's nothing much we can do about it, it's a cycle. (Dene elder, quoted in Parlee & Furgal, 2010, p. 21)

There are numerous ways that traditional knowledge holders describe the health of caribou; body condition, particularly the thickness of back fat and brisket fat, is a primary reference point for health (Moller, Berkes, Lyver, & Kislalioglu, 2004; Lyver, 2002). In addition to "fat caribou," further body condition indicators are framed as "good-looking" by the late elder J.B. Rabesca of Łutsël K'e:

> Good-looking caribou – their horns look nice and their fur is pretty white. By that you know the caribou is fat ... during the [late] fall you don't shoot the male caribou because they are skinny. They don't eat at that time because [of the rut] – they are chasing the female caribou ... When you shoot a caribou, the first thing you do is check if the caribou is fat by cutting in the middle of the stomach. If the caribou is fat, the hunter is happy. (J.B. Rabesca quoted in Parlee, Manseau, & Łutsël K'e Dene First Nation, 2005, p. 32)

Traditional knowledge suggests that the body condition of barren-ground caribou varies among animals and between years, and differences in body condition are noted with age of the caribou (Lyver, 2005).

In a study in Łutsël K'e, researchers examined the relationship between population dynamics and traditional knowledge metrics of body condition (Lyver, 2005; Lyver & Gunn, 2004). The same metrics were also documented as important to Dene and Gwich'in hunters in other parts of the Northwest Territories, the Yukon, and Alaska (Kofinas et al., 2004). Some traditional knowledge studies have strongly highlighted the importance of caribou fat as an indicator of the quality of habitat, in addition to its historic and contemporary significance to the diet of Dene, Métis, and Inuit peoples. Fat is used as an ecological indicator of caribou health and can be a signal of change in habitat condition. For example, many hunters in Łutsël K'e use caribou fat to explain habitat conditions in regions beyond their immediate harvesting areas (Lyver, 2005). According to studies on habitat and body condition of caribou in the Tłįchǫ and Łutsël K'e Dene regions, the health of the caribou is determined by several aspects of the habitat, with rain on snow being particularly problematic, as it creates a crust on the snow that limits the ability of caribou to access their food (ibid.). In contrast, rain during warm weather creates the best food (Legat, Chocolate, & Chocolate, 2008).

CONTEMPORARY INUVIALUIT, GWICH'IN, AND SAHTÚ PERSPECTIVES

Traditional knowledge about the Cape Bathurst–Tuk Peninsula, Bluenose East, Bluenose West, and Porcupine herds is available for the 2008–10 period based on presentations during community hearings as well as on interviews done with hunters in the communities of Paulatuk, Tuktoyaktuk, Fort Good Hope, Fort McPherson, and Deline. The most detailed information comes from two surveys in the Inuvialuit Settlement Region and in Fort McPherson (Gwich'in Settlement Area) that speak specifically to health and population.

In Tuktoyaktuk, twenty-four active hunters were interviewed between 2008 and 2010. In those years, the average annual harvest per hunter interviewed was 3.9 animals. Their harvest would have included Cape Bathurst and/or Tuk Peninsula caribou, which are considered to include feral reindeer from the region. Most hunters did not report getting any caribou in the spring and summer, as they did not hunt during those months.

Based on evidence from community interviews, these caribou have been abundant and healthy since the late 1960s and early 1970s. For example, in Chapter 1 of this volume, Frank Pokiak explains that the Cape Bathurst herd regularly moves back and forth between the Paulatuk area and the Tuktoyaktuk area in thirty- to forty-year cycles. He recalls the first time he hunted caribou:

> The first time I went hunting caribou was in 1967. My brother-in-law sent for us boys to go to Aklavik to hunt caribou with him. He said that there were a lot of caribou in the foothills not too far from Aklavik. When

we went out hunting, we went eight kilometres upriver from Aklavik. We
travelled from there for about one and a half hours and eventually ran into
a herd of about 150 caribou. We got six caribou that trip. That was the very
first time I hunted caribou.

During the 2008–10 period, well over half (67 percent) of the twenty-
four hunters interviewed said the caribou were in good or very good
condition. Most participants (52 percent) agreed that there had been a
change in caribou population numbers or availability of caribou in recent
years. Conversely, 13 percent reported that there had not been a change,
18 percent said that they "did not know," and 17 percent answered that
they "didn't really notice."

Residents of Paulatuk speak to changes in the range of the Bluenose
West herd. According to their harvest, the caribou have been close to
the community and abundant in the past five years. Prior to that period,
however, the caribou had left or were not around (see Chapter 9). This is
the same herd being harvested by residents of Colville Lake. Residents of
that community have not observed any changes in population; however,
the caribou are "getting harder to find": "I don't believe that the caribou's
declining; it's just that they're getting harder to find" (quoted in SRRB,
2007, p. 62).

In the Sahtú region, elders at Colville Lake describe how Bluenose
West caribou moved away from the community or "disappeared" due to
disrespect in the 1940s but have been around the community ever since
and have stayed in that same "pattern":

> Back in 1942, 1943 from back that time there were five (5) years or four
> (4) years that were – there was no caribou because of a child that had hit a
> caribou with a stick and for four (4) years there were no caribou after that.
> And after that the caribou start moving back and to this day the – the pat-
> terns have been – remained the same. And it's still the same. (Quoted in
> SRRB, 2007, pp. 113–14)

In Fort Good Hope the numbers of caribou currently available are limited.
According to some harvesters, caribou are not as abundant as they used
to be – they are getting harder to find in the region. Although hunting in
the corridor between Fort Good Hope and Colville Lake is common for
some hunters, many people have switched their focus to hunting more
moose, with community hunts for moose, Dall sheep, and mountain
caribou playing an increasingly important role in the community (see

Chapter 7). Lack of access to large numbers of barren-ground caribou in the Fort Good Hope region, which sits on the margins of the total range identified through the Government of the Northwest Territories, indicates some range shift, contraction, and decrease in population. Many elders assert that the change can be attributed to mixing between the Bluenose West and Bluenose East herds. As noted by Pokiak in Chapter 1, these are not considered one herd by the people of the region. The interpretation of "herd" dynamics seems consistent with recent aerial survey data that show a decline in the Bluenose West herd but an increase in the Bluenose East herd.

The supposition of traditional knowledge holders that the Bluenose are a single herd, despite going against the pervasive tendency of biologists to fragment and draw boundaries around smaller and smaller numbers of animals, is consistent with traditional beliefs about caribou all being the same. The exception in the Northwest Territories may be the Porcupine herd. The Gwich'in assert there is no mixing of the Porcupine and Bluenose West herds. "It is well known that Vadzaih (caribou) from the Porcupine herd never cross to the east of the Mackenzie River, nor do Vadzaih from the Bluenose ever cross to the west side" (GRRB, 1997, p. 20).

A harvester at Fort McPherson noted changes in the range of caribou during the fall of 1975, when the Porcupine caribou were very close to the community, suggesting a possible peak period of range expansion and population (Berger, 1977). Few incidences of caribou coming close to the community have been recorded since the 1970s, suggesting range has shifted or contracted. One of the core areas of caribou movement that has changed little is that of James Creek and the Vittrekwa River. As noted in the Gwich'in Land Use Plan, Dachan Dha'aii Njik (James Creek) is among the most important areas in the Gwich'in Settlement Area for Porcupine caribou migrating during the fall, spring, and winter (GRRB, 1997).

In 2008, research with twenty-seven active harvesters focused on the population, distribution, and condition of the Porcupine herd. Sixteen suggested that there was some change, six said that they had not noticed caribou population change, and four said that they did not think the caribou population had changed in recent years. All of those who perceived a change in the population said it had decreased, except one respondent who perceived the population to be growing. Population change was attributed to a variety of factors, including hunting, climate change, predation, and various kinds of disturbance, including resource development, tourism, pollutants and contaminants in the land, and natural population variability. Twenty-one of the harvesters noticed a recent change in caribou distribution, which they attributed to changes in habitat (i.e., food availability). Twenty-five hunters at Fort McPherson were also asked about their perceptions of caribou health, and twenty-four said that the caribou were in either good or very good health (Wray, 2011, p. 68). Given the ambiguous reports of caribou availability and the positive reports of condition, it is interpretable that a decline in the population of caribou was not experienced in this region. However, farther west in the range in Old Crow, a more pronounced decline was perceived by some harvesters.

In all cases, there were changes in the availability of caribou but no noted concerns about body condition or health. Caribou were still considered fat or in good condition. Although the data are preliminary, the absence of change in body condition suggests that no major crash in population has occurred. Such data may also affirm the supposition being put forward by hunters in many parts of the western Arctic that the decrease in caribou availability can be attributed to a shift in the range rather than a decline in population.

* * *

The uncertainty of caribou comings and goings is a well-accepted characteristic of northern livelihoods (Nuttall et al., 2005; Smith, 1978). Variabilities in the comings and goings of caribou can be understood from many different positions. The dominant paradigm and methodology of caribou management in the Northwest Territories has centred on the arithmetic of caribou counting. Done through aerial surveys that are sometimes based on tenuous methodological and ecological assumptions, these caribou counts are often the focus of public scrutiny and sometimes litigation. A key concern in many Aboriginal communities is that despite the margins of error, counts are used uncritically by governments to impose drastic harvest regulations that have significant socio-cultural and health implications. "Harvest" is an easy scapegoat; a large part of society, particularly urban society, easily dismisses or demonizes hunting as an unnecessary act of violence. Resource development arguably represents another kind of ecological violence but one that is more digestible or politically and economically convenient. Since the precautionary principle seems liberally applied to subsistence livelihoods, communities question why it is not applied more fairly in decision making about other known stressors on caribou and caribou habitat, such as oil and gas and mining activity. During the same period that territorial governments and co-management boards imposed harvest restrictions on communities dependent on the Porcupine, Bluenose, and Cape Bathurst–Tuk Peninsula herds, the Yukon and Northwest Territories maintained open-door policies for resource development. In the Porcupine range alone, mining exploration is increasing exponentially. The Sahtú region has also seen unprecedented growth in gas exploration. Coupled with climate change, these issues may indeed have far greater implications for caribou in the region – "people play a great role in shaping landscapes and thus indirectly affect the size and nature of Rangifer populations" (Anderson, 2000, p. 154).

Although much criticism is laid at the doorstep of scientists and science, the problem may be in the poor interpretation of science. Governments and media often position scientific hypotheses about caribou numbers as concrete realities, and regulations are made as though no uncertainties existed at all. When numbers prove to be inaccurate, the credibility of science comes into question, even though the culprits are in fact biased interpretation on the part of politicians and crisis-oriented reporting. The scientists who are blamed when the predictions are proven inaccurate may suffer longer-term consequences as caribou users begin to have less faith and trust in "caribou science." The role of traditional knowledge in this landscape of "caribou science" has thus become important.

REFERENCES

Aanes, R., Saether, B.E., Smith, F.M., Cooper, E.J., Wookey, P.A., & Oritsland, N.A. (2002). The Arctic Oscillation predicts effects of climate change in two trophic levels in a High-Arctic ecosystem. *Ecology Letters*, 5(3), 445–453.

Albon, S.D., Stien, A., Irvine, R.J., Langvatn, R., Ropstad, E., & Halvorsen, O. (2002). The role of parasites in the dynamics of a reindeer population. *Biological Sciences*, 269(1500), 1625–1632.

Anderson, D.G. (2000). *Identity and Ecology in Siberia: The Number One Reindeer Brigade*. Oxford: Oxford University Press.

Berger, T. (1977). Fort McPherson, NT – Berger Commission reports and community transcripts (July 18, 1975). In *Northern Frontier, Northern Homeland: The Report of the Mackenzie Valley Pipeline Inquiry*. Ottawa: Minister of Supply and Services Canada.

Bergerud, A.T. (1996). Evolving perspectives on caribou population dynamics, have we got it right yet? *Rangifer*, 16(9), 95–116.

Bergerud, A.T., Jakimchuk, R.D., & Carruthers, D.R. (1984). The buffalo of the North: Caribou (*Rangifer tarandus*) and human development. *Arctic*, 37(1), 7–22.

Blondin, G., & Blondin, J. (2009). *The Legend of the Caribou Boy*. Penticton, BC: Theytus.

Boulanger, J., Gunn, A., Adamczewski, J., & Croft, B. (2011). A data-driven demographic model to explore the decline of the Bathurst caribou herd. *Journal of Wildlife Management*, 75(4), 883–896.

Boulanger, J., Poole, K.G., Gunn, A., & Wierzchowski, J. (2012). Estimating the zone of influence of industrial developments on wildlife: A migratory caribou *Rangifer tarandus groenlandicus* and diamond mine case study. *Wildlife Biology, 18*(2), 164–179.

Brotton, J., & Wall, G. (1997). Climate change and the Bathurst caribou herd in the Northwest Territories. *Climatic Change, 35*(1), 35–52.

Cameron, R.D., Lenart, E.A., Reed, D.J., Whitten, K.R., & Smith, W.T. (1995). Abundance and movements of caribou in the oilfield complex near Prudhoe Bay, Alaska. *Rangifer, 15*(1), 3–8.

Crete, M., & Payette, S. (1990). Climatic changes and caribou abundance in northern Quebec over the last century. *Rangifer, 10*(3), 159–165.

Ferguson, M.A., Gauthier, L., & Messier, F. (2001). Range shift and winter foraging ecology of a population of Arctic tundra caribou. *Canadian Journal of Zoology, 79*(5), 746–758.

Gunn, A., Russell, D., White, R., & Kofinas, G. (2009). Facing a future of change: Wild migratory caribou and reindeer. *Arctic, 62*(3), iii–iv.

Gwich'in Renewable Resources Board (GRRB). (1997). *Nành' Kak Geenjit Gwich'in Ginjik (Gwich'in Words about the Land)*. Inuvik: Gwich'in Renewable Resources Board.

Huebert, R. (2011). Canadian Arctic sovreignty and security in a transforming circumpolar world. In F. Griffiths, R. Huebert, & P.W. Lackenbauer (Eds.), *Canada and the Changing Arctic: Sovereignty, Security, and Stewardship* (pp. 13–59). Waterloo, ON: Wilfrid Laurier University Press.

Johnson, C.J., Boyce, M.S., Case, R.L., Cluff, H.D., Gau, R.J., Gunn, A., & Mulders, R. (2005). Cumulative effects of human developments on Arctic wildlife. *Wildlife Monographs*, (160), 1–36.

Kelsall, J.P. (1968). *The Migratory Barren-Ground Caribou of Canada*. Ottawa: Indigenous and Northern Affairs Canada and Canadian Wildlife Service.

Kendrick, A. (2002). Caribou co-management: Realizing conceptual differences. *Rangifer, 22*(4), 7–13.

Kofinas, G., Lyver, P.O., Russell, D., White, R., Nelson, A., & Flanders, N. (2004). Towards a protocol for monitoring of caribou body condition. *Rangifer, 23*(14), 43–52.

Kuhnlein, H.V., Receveur, O., Soueida, R., & Egeland, G.M. (2004). Arctic indigenous peoples experience the nutrition transition with changing dietary patterns and obesity. *Journal of Nutrition, 134*(6), 1447–1453.

Kutz, S., Elkin, B., Panayi, T., & Dubey, J.P. (2001). Prevalence of *Toxoplasma gondii* antibodies in barren-ground caribou (*Rangifer tarandus groenlandicus*) from the Canadian Arctic. *Journal of Parasitology, 87*(2), 439–442.

Legat, A. (2012). *Walking the Land, Feeding the Fire: Knowledge and Stewardship among the Tłįchǫ Dene*. Tucson: University of Arizona Press.

Legat, A., Chocolate, G., & Chocolate, M. (2008). *Monitoring the Relationship between People and Caribou: Tłįchǫ Laws and Indicators of Change*. Yellowknife: West Kitikmeot Slave Study Society.

Legat, A., Chocolate, G., Gon, B., Zoe, S.A., & Chocolate, M. (2001). *Caribou Migration and the State of Their Habitat*. Yellowknife: West Kitikmeot Slave Study Society.

Lyver, P.O. (2002). *Use of First Nations Dene Knowledge to Monitor Changes in Barren Ground (Rangifer tarandus groenlandicus) Body Condition*. Winnipeg: University of Manitoba Press.

Lyver, P.O. (2005). Monitoring barren-ground caribou body condition with Denésǫłıné traditional knowledge. *Arctic*, *58*(1), 44–54.

Lyver, P.O., & Gunn, A. (2004). Calibration of hunters' impressions with female caribou body condition indices to predict probability of pregnancy. *Arctic*, *57*(3), 233–41.

Messier, F., Huot, J., Le Henaff, D., & Luttich, S. (1988). Demography of the George River caribou herd: Evidence of population regulation by forage exploitation and range expansion. *Arctic*, *41*(4), 279–87.

Moller, H., Berkes, F., Lyver, P.O., & Kislalioglu, M. (2004). Combining science and traditional ecological knowledge: Monitoring populations for co-management. *Ecology and Society*, *9*(3), 2. Retrieved from http://www.ecologyandsociety.org/vol9/iss3/art2

Nadasdy, P. (2005). The anti-politics of TEK: The institutionalization of co-management discourse and practice. *Anthropologica*, *47*(2), 215–232.

Nuttall, M., Berkes, F., Forbes, B., Kofinas, G., Vlassova, T., & Wenzel, G. (2005). Hunting, herding, fishing, and gathering: Indigenous peoples and renewable resource use in the Arctic. In C. Symon, L. Arris, & B. Heal (Eds.), *Arctic Climate Impact Assessment* (pp. 649–690). New York: Cambridge University Press.

Parlee, B., Basil, M., & Drybones, N. (2001). *Traditional Ecological Knowledge in the Kache Kue Study Region: Final Report*. Yellowknife: West Kitikmeot Slave Study Society.

Parlee, B., & Furgal, C. (2010). Communities and caribou workshop summary report. In *Arctic Peoples, Culture, Resilience and Caribou*. Edmonton: University of Alberta Press.

Parlee, B., Manseau, M., & Łutsël K'e Dene First Nation. (2005). Using traditional knowledge to adapt to ecological change: Denésǫłıné monitoring of caribou movements. *Arctic*, *58*(1), 26–37.

Payette, S., Boudreau, S., Morneau, C., & Pitre, N. (2004). Long-term interactions between migratory caribou, wildfires and Nunavik hunters inferred from tree rings. *Ambio*, *33*(8), 482–486.

Porcupine Caribou Management Board (PCMB) (2011). *Counting the Caribou*. Whitehorse: Porcupine Caribou Management Board. https://www.pcmb.ca/PDF/researchers/Herd-Status/Counting%20the%20Caribou%20Brochure%20-%20final.pdf.

Riley, S.J., Decker, D.J., Carpenter, L.H., Organ, J.F., Siemer, W.F., Mattfeld, G.F., & Parsons, G. (2002). The essence of wildlife management. *Wildlife Society Bulletin*, *30*(2), 585–593.

Rinfret, S. (2009). Controlling animals: Power, Foucault, and species management. *Society & Natural Resources*, *22*(6), 571–578.

Romesburg, H.C. (1981). Wildlife science: Gaining reliable information. *Journal of Wildlife Management*, *45*(2), 293–313.

Russell, D.E., van de Wetering, D., White, R.G., & Gerhart, K.L. (1996). Oil and the Porcupine caribou herd – Can we quantify the impacts? *Rangifer*, *16*(9), 255–258.

Ruttan, R. (2012). New caribou crisis – Then and now. *Rangifer*, *32*(20), 85–102.

Sahtú Renewable Resources Board (SRRB) (2007). *Public Hearing: Bluenose-West Management Hearing*. Fort Good Hope: Sahtú Renewable Resources Board. http://www.srrb.nt.ca/index.php?option=com_content&view=category&id=142&Itemid=1225.

Sandlos, J. (2004). *Northern wildlife, northern people: Native hunters and wildlife conservation in the Northwest Territories* (PhD diss.). York University, Toronto.

Smith, D.M. (2002). The flesh and the word: Stories and other gifts of the animals in Chipewyan cosmology. *Anthropology and Humanism*, *27*(1), 60–79.

Smith, J.G.E. (1978). Economic uncertainty in an "original affluent society": Caribou and caribou eater Chipewyan adaptive strategies. *Arctic Anthropology, 15*(1), 68–88.

Spak, S. (2005). The position of Indigenous knowledge in Canadian co-management organizations. *Anthropologica, 47*(2), 233–246.

Thomas, D. (1996). Needed: Less counting of caribou and more ecology. *Rangifer, 18*(10), 15–23.

Thomas, L., Buckland, S.T., Newman, K.B., & Harwood, J. (2005). A unified framework for modelling wildlife population dynamics. *Australian & New Zealand Journal of Statistics, 47*(1), 19–34.

Thorpe, N., Hakongak, N., Eyegetok, S., & Kadlun-Jones, M. (2001). *The Tuktu and Nogak Project Final Report: A Caribou Chronicle.* Yellowknife: Government of the Northwest Territories.

Wray, K. (2011). *Ways we respect caribou: Hunting in Teetł'it Zheh (Fort McPherson, NWT)* (MSc thesis). University of Alberta, Edmonton.

4

Beyond the Harvest Study

Brenda Parlee, Natalie Zimmer,
and Peter Boxall

Harvest studies have been an important tool for quantifying Aboriginal use of wildlife resources in the Canadian North for many decades. "Native harvest studies" date back to the James Bay and Northern Quebec Agreement of 1976 and have been more recently included as part of the settlement of land claims in the region of Nunatsiavut and elsewhere (Berkes, 1983; Usher, 2002; Usher & Wenzel, 1987). Harvest data from the Inuvialuit and Gwich'in regions of the Northwest Territories were analyzed as part of this study to learn more about the patterns of harvest of barren-ground caribou (i.e., Bluenose West, Porcupine, and Cape Bathurst). Given that the populations of the three barren-ground herds were reportedly in decline during the period of the harvest study (1999–2003), we sought to determine whether the harvest of caribou also declined. Although data on harvest "effort" were not available for the Inuvialuit and Gwich'in regions, we hypothesize that harvest numbers remained constant. This hypothesis is based on preliminary analysis of the Sahtú harvest data, which show that harvesting effort remained relatively constant, and of other socio-economic data from the Inuvialuit and Gwich'in communities.

HARVESTING AND SENSITIVITY TO RESOURCE AVAILABILITY

The use of harvest data to calculate or track changes in wildlife populations is common practice in many parts of the world. It is considered by some scholars one key way that Indigenous knowledge can inform wildlife management (Gilchrist, Mallory, & Merkel, 2005). Scholars have argued that the number of animals harvested is a measure embedded in many Indigenous resource management systems; "the harvest rate, or similar catch per unit of effort (CPUE) measurement, is the most practical population-monitoring index for customary resource users" (Moller et al., 2004, p. 1). In the absence of other kinds of population surveys such as caribou counts, CPUE is thought to be a useful proxy for understanding population dynamics; the greater the population, the less effort is required in harvest and vice versa. "The use of CPUE as an index of abundance rests on the assumption that catch is proportional to both the abundance of the harvested population and the amount of effort invested in hunting" (Keane, Jones, & Milner-Gulland, 2011, p. 1165). However, simple summing of harvest outcomes and CPUE analysis are not the same thing. Harvest studies tend to focus only on the number of animals taken, with less consideration given to the various aspects of "effort," including input and opportunity costs, distances, and time. CPUE also assumes the recording of both successful harvest events and unsuccessful events or trips. Unsuccessful trips are sometimes not well documented during harvest studies due to a lack of recall by respondents and/or the overarching emphasis on calculating numbers of retrieved animals.

Economic anthropologists and resource economists have theorized that harvesters generally adapt to the declining availability of wildlife and other valued resources in the short term by increasing the level of effort to find those resources (Winterhalder, 1981). This notion of optimal foraging suggests that there is a tipping point at which it no longer becomes cost effective to harvest the same species, in the same way, or in the same places. Harvesters must then decide, given the lack of harvest success, whether to continue along the same path or to make either small or significant changes to their harvesting practices (Moran, 1982). At that point, and in the context of longer-term scarcity of resources, harvesters are likely to make greater changes in their livelihood practices.

HARVEST AND DIET

Caribou comprise an important component of traditional food economies as well as diets in northern Canada. It is well established that traditional food such as caribou meat is nutritionally superior to many other kinds of store-bought foods known to be higher in carbohydrates and fats (Chiu,

Goddard, & Parlee, 2016; Kuhnlein et al., 2013; Receveur, Boulay, & Kuhnlein, 1997). The proportion of traditional food in the total diet of northern communities varies, with communities that are more northerly consuming a greater proportion of country traditional foods than those to the south.

> Differences in traditional food intake among communities may be related to community characteristics such as population size, road access and availability of affordable market food, proximity to animal migration routes, and prevalent fishing and hunting practices. Of particular interest is how differences in traditional food use are likely to be reflected in differences in diet quality and nutrient intakes. (Receveur, Boulay, & Kuhnlein, 1997, p. 2183)

The continuance of traditional diets is considered to be protective against the onset of many kinds of lifestyle diseases, including diabetes, some cancers, and chronic heart disease. Conversely, across many parts of the North, the pervasive decline in the consumption of traditional foods and the transition to "American" diets that are higher in fat and carbohydrates have been linked to the increasing prevalence of such lifestyle diseases. In the 1990s, caribou and moose were the terrestrial species most commonly consumed in northern communities (Kuhnlein & Receveur, 2007, p. 1111). Based on data from the Inuvialuit harvest study, the caribou meat available from harvest amounted to 110.37 kilograms per year, accounting for 33 percent of the total traditional harvest in Inuvialuit communities (Usher, 2002, p. 23). In the Gwich'in region, the caribou meat available from harvest was 553.91 kilograms over five years, accounting for more than 75 percent of the total traditional food harvest (GRRB, 1997). In the Sahtú region, harvest numbers are roughly calculated to be comparable in aggregate to those of the Gwich'in region but vary by community (McMillan, 2011).

The replacement value of caribou meat – a measure often used by economists and food security analysts – has been estimated to be in the tens of millions of dollars in the Northwest Territories as a whole and to be hundreds of thousands of dollars per year in most of the Inuvialuit, Gwich'in, and Sahtú communities (Ashley 2002). The proportion of caribou represented in livelihood and diet is not the ideal measure of significance; the spiritual and cultural significance goes well beyond these narrow measures. Within that context, the relative decline in harvest and consumption of caribou meat thus has significant human health implications unless there is harvest substitution or a relative increase in the harvest and consumption of other kinds of traditional foods or nutritionally valuable store-bought foods.

HARVEST SUBSTITUTION

"Arctic hunters and herders have always lived with, and adapted to, shifts and changes in the size, distribution, range and availability of animal populations. They have dealt with flux and change by developing significant flexibility in resource procurement techniques and in social organization" (Nuttall et al., 2005, pp. 11–27). In the Gwich'in and Inuvialuit regions, as in many parts of northern Canada, traditional livelihoods are not specialized; although caribou represent a significant proportion of the wild meat consumed during peak population years; moose and a diversity of fish species, marine mammals, migratory geese and ducks, and small game also contribute to household diets (Kuhnlein & Receveur, 2007; Usher, 2000). Averages are thus deceiving since what is harvested and shared on the dinner table varies significantly from season to season and year to year. The availability of each species will affect harvest and consumption patterns over time.

How do harvesters cope with declines in the availability of caribou? For some, the choices to invest in additional fuel and travel time in order to reach the far edges of the caribou range (i.e., to search for moose instead of caribou) or to set additional fishnets or rabbit snares are not random. Bousman (1993, p. 65) suggests that communities employ strategy in their harvesting by balancing their focus on high-risk species characterized by significant variability and uncertainty with the harvest of resources of greater predictability:

> In a dynamic situation hunter-gatherers can shift from a risk-prone to a risk-averse strategy or the reverse as the availability of resources fluctuates through time. With a diversified resource base provided by a gender-based division of labor and intra-group sharing, hunter-gatherers often balance the risk-averse exploitation of predictable resources (usually plants or fish) against the exploitation of unpredictable resources (usually terrestrial mammals) by risk-prone strategies.

For many people, harvest substitution is not a theoretical or economic decision but a seamless response to changes in recognized patterns of resource availability and one of the key reasons behind their respect for ecological diversity.

Socio-economic Factors Affecting Harvest

The mixed economy is a well-established lens through which anthropologists, geographers, economists, and others have examined the interconnections between the subsistence and wage-based economies (Usher, Duhaime, & Searles, 2003). In many regions, the wage economy is an acknowledged driver of declining participation in hunting, trapping, and fishing, a trend precipitated and compounded by the rapid pace and scale of socio-cultural change across the North (Poppel et al., 2007; Usher, Duhaime, & Searles, 2003). Participation in the wage economy varies; the reported or formal unemployment rates can be as high as 50 percent in some communities in the Gwich'in, Inuvialuit, and Sahtú regions in some years and closer to the national Canadian average in others (e.g., 7 percent). For some people, very low incomes or no wage employment opportunities may result in more time for harvesting, particularly in the short term (see Chapter 9). For harvesters, wages make it possible to purchase the necessary equipment for harvest, which may include snowmobiles, rifles, and camping supplies. In the long term and for people with middle to higher incomes, increased participation in wage employment may translate into a busy work schedule, less time for other activities, and diminished social networks, with income leading to no added benefit.

Harvest Study Methods

Harvest studies have many stated objectives and mandates tied to Aboriginal rights enshrined in treaties and in comprehensive land claim agreements like that of the Gwich'in – notably the right to "gather, hunt, trap and fish throughout the settlement area at all seasons of the year" (INAC, 1992). The collection, management, and use of harvest data have been complicated by numerous factors (Usher & Wenzel, 1987). Among the socio-political issues affecting both community participation and perception of the value of such studies is the use of data in wildlife harvest management (BCMPWG, 2011; Gunn et al., 2011).

The position of harvest statistics in wildlife management is tenuous, particularly during periods of resource scarcity, such as the reported period of declining caribou populations. Academic, federal, and territorial authorities and co-management boards have been perennially preoccupied with the impact of Aboriginal harvesting on natural resources (Sandlos, 2004). Government support for harvest studies in the Northwest Territories has been grounded in concerns about a potential "tragedy of the commons" scenario; those who take this view suggest that overexploitation of barren-ground caribou is inevitable unless the state intervenes to ensure sustainable harvest (BCMPWG, 2011; Kruse et al., 2004). At the same time, some Aboriginal governments have valued harvest studies for their ability to highlight the numerical significance of northern resources to local livelihoods and thus have regarded them as tools to further land claims and affirm treaty rights, which have been legally interpreted as hinging largely on the traditional practices of hunting, trapping, and fishing. Some might say that harvest studies are a necessary "evil" – a means of representing Aboriginal interests in lands and resources within a political economy dominated by quantitative valuation. In that context, they are both insurance and an assurance against potential losses from resource development (e.g., mining, forestry, and oil and gas development), where the calculation of replacement value (based on edible weights) demands detailed data on resource dependence (Haener, Dosman, & Boxall, 2001). Nonetheless, there are concerns within communities and among some scholars and Aboriginal leaders that the quantitative methods used in harvest studies cannot account for the diverse social, cultural, and spiritual values inherent in caribou and other wildlife (Berkes, 2008). The disconnect between the mandate of harvest studies, as established in land claim agreements, and their end use in harvest management has arguably contributed to a distrust of harvest statistics on the part of Aboriginal communities, particularly during periods of resource scarcity.

Opportunities for harvest surveys in the Inuvialuit, Gwich'in, and Sahtú regions were created with land claim agreements in those regions in the mid-1980s and early 1990s. The common objective of these studies, each five to six years in duration, was the identification of a baseline of resource dependence, or a "minimum needs level." These studies, and indeed the majority of harvest studies throughout Canada, have focused consistently on five parameters: (1) harvester profile, (2) species categories, (3) data of harvest event, (4) locational data (i.e., area), and (5) quantity of species harvested (Berkes, 1990; Filion, 1980; Usher & Wenzel, 1987). Most harvest studies show a primary concern with what is sometimes called "strike and retrieve" information (Usher & Wenzel, 1987,

p. 148), including biological details (e.g., species and genus) and locations of harvest. Additional attention also seems to be given in most studies to the Aboriginal status (i.e., First Nation, Inuit, or Métis) of the harvester, with the apparent aim being to surveil his or her resource use (Usher & Wenzel, 1987, p. 146).

The largest gap in data collection has been in details about each hunter as a socio-economic actor who makes harvest decisions based on a complicated set of factors, such as time, income, access to other sources of food, and so on. This gap may be tied to the preponderance of scholars and others who homogenize their view of Aboriginal communities or assume that all have the same values, employment status, and education (Agrawal & Gibson, 1999; Natcher & Hickey, 2002). Lacking details about the diversity of beliefs, attitudes, choices, and behaviours, harvest studies, like many kinds of academic investigations *about* Aboriginal communities, have arguably reproduced simplifications of cultures, economies, and human-environment relations in northern Canada and elsewhere.

Drawing upon the harvest study data from the Inuvialuit and Gwich'in communities, an analysis of the following questions was carried out. What is the relative importance of caribou within the total harvest of the Inuvialuit and Gwich'in communities? To what extent are declines in the harvest of caribou offset by the harvest of other species? What ecological and socio-economic factors might explain the variation in barren-ground caribou harvests?

METHODS

The analysis of harvest data presented here is based on publicly available harvest data from the Inuvialuit Game Council and the Gwich'in Renewable Resources Board (McDonald & GRRB, 2009; Inuvialuit Joint Secretariat, 2003). The Inuvialuit and Gwich'in harvest studies were established under the terms of their respective land claim agreements and ran from 1986 to 1997 in the Inuvialuit region and from 1995 to 2001 in the Gwich'in region. A similar study was also carried out from 1998 to 2005 in the Sahtú region; however, no data were available from this study for this volume.

As with other harvest studies, the objectives were to provide information on harvesting necessary for the effective management of fish and wildlife in this region and to determine a "minimum needs level" – or the proportion of total allowable harvest of the species required to meet and sustain the needs of Gwich'in and Inuvialuit beneficiaries. The issue is controversial for many communities in that it does not necessarily take into consideration the natural ups and downs of population dynamics beyond the life of the study. It also assumes "need" statistically rather than normatively. What is considered a need by one community member may be different from what is considered a need by others. Moreover, need is influenced by aspects of the livelihood that may change over time, such as wage employment and the availability of other species. Despite these limitations, the harvest study data provide a useful perspective on the significance of caribou and other resources to the diets and economies of northern communities.

A list of active harvesters was developed that formed the foundation of data collection efforts for the course of both the Gwich'in and Inuvialuit studies (McDonald & GRRB, 2009; Inuvialuit Joint Secretariat, 2003). Community researchers were hired and trained for each of the study years. Monthly recall questions were designed according to a regional fish harvest calendar; the intent was to account for all fish and wildlife species harvested. The total species list for each community varied from a low of seventeen to over sixty. The survey instrument was otherwise similar for each community and included (1) harvester age and gender, (2) species (e.g., wildlife, fish, geese, ducks), (3) quantity of species harvested, (4) time of harvest, and (5) place of harvest (i.e., gridded map sheets at a scale of 1:250,000). Participation was considered good, with a response rate of 80–95 percent. Response rates were essentially calculated as the number of hunters interviewed divided by the total number of hunters multiplied by 100 percent; to account for misreporting, a margin of error of 5 percent was included. A coefficient of variance was also calculated, adding to the margin of error in some months and some communities.

BARREN-GROUND HARVESTING TRENDS

The harvest of barren-ground caribou was documented for 1988–97 in the Inuvialuit Settlement Region and for 1995–2000 in the Gwich'in Settlement Area. Harvest numbers were highly variable and in overall decline during the study periods (see Figures 4.1, 4.2, and 4.3); this pattern can be attributed to a variety of factors.

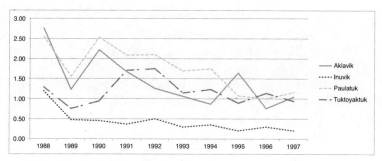

FIGURE 4.1   Inuvialuit harvest of caribou (per capita) by community, 1988–97
Source: Based on Inuvialuit Joint Secretariat (2003).

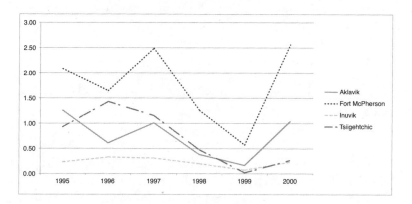

FIGURE 4.2   Gwich'in harvest of Porcupine caribou (per capita) by community, 1995–2000
Source: Based on McDonald & GRRB (2009).

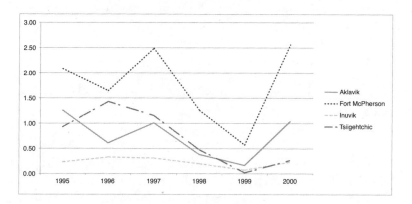

FIGURE 4.3   Gwich'in harvest of Bluenose caribou (per capita) by community, 1995–2000
Source: Based on McDonald & GRRB (2009).

Interpretations of Harvest Data

Methodological Considerations

How dependable are the harvest data reported? Methodological concerns have been raised in regard to many harvest studies in many parts of northern Canada (Berkes, 1990; Usher & Wenzel, 1987). Does the variability in harvest numbers reflect real changes in harvesting? Some degree of error is considered consistent with social surveys involving recall. Some critics might suggest that the margin of error associated with harvest studies in the North is greater than that of the average social survey due to the political nature of Aboriginal harvesting. On the flipside, harvest reporting in the Inuvialuit and Gwich'in regions may be considered more accurate and valid than the other kinds of social survey data given that the acquisition of caribou meat is a highly transparent activity in northern communities; when a hunter pulls up a sled or truck to his or her house at the end of a hunting trip, it is plain to see how many caribou have been retrieved. In other words, reporting is more likely to be accurate due to the oversight of nosy neighbours and family members.

Previous analysis of the Inuvialuit harvest data suggested that the margin of error associated with the data was insignificant (Usher, 2002, p. 19). The standard margin of error used in the analysis of both the Gwich'in and the Inuvialuit harvest data was 5 percent. However, the coefficient of variability was considered higher in some months and some communities. But even with a significant margin of error, the decline in harvest represented in the Inuvialuit and the Gwich'in harvest data is significant (52 percent). The most significant decline reported during the first years of the study was in the Aklavik harvest of Bluenose caribou (93 percent). The least overall decline was in Aklavik for Porcupine caribou, with the data suggesting a 17% percent decrease from the beginning of the study to the final year (see Table 4.1).

In addition to the seasonal and year-to-year variability in populations and distribution, hunting trips can vary widely as a result of a number of factors, including climate, transportation (e.g., community hunts), input costs (e.g., fuel), and opportunity costs associated with other livelihood pursuits. Although climate change presents a major stress on harvesters in some regions and communities, no trends were noted in an analysis of regional data on temperature and precipitation that would explain the harvest variability and decline. Similarly, fuel prices remained relatively constant during the years of the harvest studies in the Inuvialuit and

TABLE 4.1   Percentage change in harvest over the study period

Communities	Change in harvest (%)
Caribou	
Aklavik	−62
Inuvik	−82
Paulatuk	−54
Tuktoyaktuk	−27
Bluenose caribou	
Aklavik	−93
Fort McPherson	n/a
Inuvik	−84
Tsiigehtchic	−60
Porcupine caribou	
Aklavik	−17
Fort McPherson	−22
Inuvik	0
Tsiigehtchic	−71

Gwich'in regions. Given the theoretical focus on the wage economy as a driver behind traditional economy declines, some data on harvest effort and labour participation are discussed.

Socio-economic Considerations: Harvest Effort and Labour Participation

Data on catches per unit of effort and theory from other studies suggest that as the population of a given species declines, harvest effort is likely to increase in order to ensure the same harvest yield. Conversely, if harvest effort remains the same during a period of population decline, one might anticipate seeing a decline in harvest yield. In the absence of data on harvest effort, some scholars suggest that data on harvest yield divided by harvest effort indicators of distance and time can provide a fuzzy indication of increases or decreases in population. Using the level of harvest effort for barren-ground caribou is more challenging than for other ungulates such as moose or deer.

To sustain the same yield, harvesters located outside or on the periphery of the range may have to increase effort in order to ensure sustained harvest yields, whereas other communities located centrally within the range contraction would not need to increase their effort. In cases where there is increased aggregation, as can occur during population declines, the harvest effort may even decrease in some communities.

In a preliminary analysis of the Sahtú harvest data, harvest effort did not appear to change for any of the communities reporting. However, for the Inuvialuit and Gwich'in harvest study analysis, no data on harvest effort were available that could be used to determine whether the changes in harvest reflected changes in harvest effort. In the absence of such data on harvest effort, other kinds of information can be useful – specifically labour participation.

The literature suggests that the wage economy has a tendency to draw some people away from harvesting, thus curtailing harvest effort and ultimately leading to a decline in total harvest (see Chapter 11). The labour participation rate for Inuvialuit communities remained relatively constant for the years of the harvest study in the Inuvialuit region. Similarly, the labour participation rate for the Gwich'in communities during their harvest study remained relatively constant. There were no major oscillations from year to year in the labour participation rate that could explain the ups and downs and the decline in harvesting activity.

Ecological Considerations: Availability of Caribou

If methodological considerations and harvest effort (i.e., labour participation) cannot explain the harvest data, ecological explanations are an important next step in analysis. Caribou are a highly dynamic species; even in years of population abundance, the distribution of caribou can change significantly, leading to variations in harvest. Each of the four herds was reported to be in decline during the 1987–2005 period of the harvest studies (see Figure 4.4). The Cape Bathurst herd is estimated to have peaked at just under 20,000 animals in 1992 (Davison et al., 2009) and had fallen to just under 2,300 animals by 2015 (Advisory Committee for Cooperation on Wildlife Management, 2014; Environment and Natural Resources, 2011). Estimates for the Bluenose West herd suggest the population peaked in the early 1990s at just over 110,000 animals (CARMA, n.d.) and had fallen significantly to just under 20,000 animals by 2015 (Advisory Committee for Cooperation on Wildlife Management, 2014; Environment and Natural Resources, n.d.). The same data from the Government of the Northwest

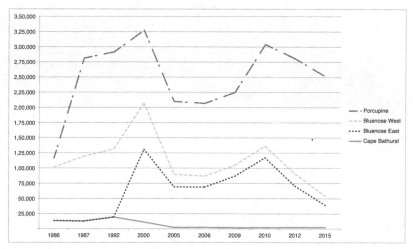

FIGURE 4.4   Population estimates for barren-ground caribou herds in the study area, 1986–2015

Source: Based on post-calving aerial survey data cited in Porcupine Caribou Management Board (n.d.); Advisory Committee for Cooperation on Wildlife Management (2014); and Environment and Natural Resources (n.d.).

Territories show the population of the Bluenose East herd to be 119,000 animals in 2000, with a decline to just under 35,000 animals by 2015. The Porcupine caribou population is thought to have peaked in 1989 at roughly 178,000 animals before falling to just over 120,000 animals in 2001, but it has since rebounded to over 190,000 animals (Porcupine Caribou Management Board, n.d.).

IMPLICATIONS

Substitution for Other Traditional Foods

In periods of caribou harvest decline, are harvesters and households substituting other species for caribou to ensure a constant availability of traditional food? Using data on the harvest per capita of barren-ground caribou and moose, there is little evidence that Gwich'in communities were substituting moose for caribou during the years of the harvest study and the years of caribou harvest decline (see Figure 4.5). A coarse analysis of the Inuvialuit harvest study similarly suggests that there is no pattern in caribou harvest decline and substitution of moose for caribou (see Figure 4.6); nor is there seemingly any such substitution of other species, including marine species.

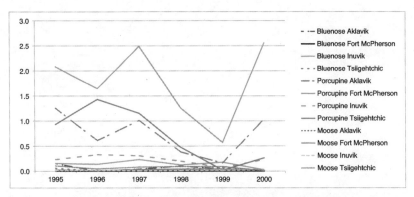

FIGURE 4.5   Gwich'in caribou and moose harvest (per capita) by community, 1995–2000
Source: Based on McDonald & GRRB (2009); and Inuvialuit Joint Secretariat (2003).

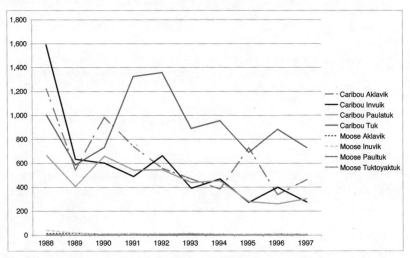

FIGURE 4.6   Inuvialuit caribou and moose harvest (per capita) by community, 1988–97
Source: Based on McDonald & GRRB (2009); and Inuvialuit Joint Secretariat (2003).

A preliminary analysis of the Sahtú harvest data also indicated no substitution of moose for caribou in this region during declining caribou harvest.

Given the lack of evidence of traditional food substitution in the harvest data, it is assumed that the substitution of store-bought food for caribou is occurring. Such a trend would be consistent with nutrition data for the same period (Receveur, Boulay, & Kuhnlein, 1997). Although this dietary transition is thought to be caused by a host of socio-cultural factors (e.g., influence of food ads on television), it may be that the decline in availability of caribou is compounding or reinforcing the pace and significance of this transition.

DISCUSSION

The harvesting practices of Inuvialuit and Gwich'in communities are highly variable, with a significant decline in all communities harvesting both Porcupine caribou and Bluenose West caribou during the study years. Although some critics of the data may suggest that methodological explanations account for the decline (e.g., fewer harvesters were reporting in the later years of the study), even with a greater margin of error, a statistically significant decline in harvesting is still in evidence. An increase in wage employment opportunities during the years of the study might offer another explanation for harvest numbers; however, using labour participation as the key indicator of community engagement in this sector, there appears to be no relationship between participation in the wage economy and the reported changes in harvesting.

Given that no other evidence or explanation has come to light, the most obvious reason for declining harvests over the study period is that harvesters began to find it harder to locate caribou during the period of the harvest studies or that harvesters became sensitive to reports of population decline. Theoretically, such responsiveness of Indigenous harvesters to resource abundance or variability is well established; based on evidence from elsewhere, it is suggested that those land-based communities with long histories of dependence on particular resources are highly adaptive (Nuttall et al., 2005). The general assumption of optimal foraging theory is that harvest yield decreases or harvesting effort increases as resource availability declines. The data on the Inuvialuit and Gwich'in regions is generally consistent with such a theory, suggesting that these communities are highly responsive to ecological conditions. There was no evidence, however, of the substitution of other traditional foods, such as moose, leading to concerns about the economic impacts of a caribou harvest decline if substitution is occurring with store-bought foods. Although more research and analysis are needed, the implication is that in the absence of caribou meat, harvesters and their households may be turning to store-bought foods.

The harvest studies in the Inuvialuit and Gwich'in regions provide an understanding of harvest patterns during the years of the study that can help to explain harvester behaviour during periods of population decline. However, to more fully explain harvester behaviour, other socio-economic data may be useful in future harvest studies (see Table 4.2).

TABLE 4.2 Socio-economic variables that might be considered in future harvest studies

Harvest effort attributes	Socio-economic factors	Environmental factors
Detailed distance/ days per harvest trip	Employment status	Species (and quantity) observed per trip
Input costs	Household labour participation	Species health observations per trip (e.g., body condition)
Harvester subsidies	Household income	Species distribution patterns
Hunting technology employed	Household size (as an assessment of possible need)	Competing land and resource use
Harvest plan versus harvest outcomes	Degree of harvest sharing within and outside the community	Environmental disturbances (e.g., tailing pond spills, pipeline leaks)

* * *

Harvest studies across northern Canada are thought to have made a major contribution to social science knowledge about harvest practices in the North and are substantially more rigorous than historical harvest estimates, which were the predominate means of tracking harvests until the 1970s. The results of this analysis suggest that harvesters in the Inuvialuit and Gwich'in regions were responsive to the declining availability of caribou during the years of the study; however, more research is needed to explore the harvest data within the context of other kinds of socio-cultural and economic change.

REFERENCES

Advisory Committee for Cooperation on Wildlife Management (2014). *Taking Care of Caribou: Cape Bathurst, Bluenose-West, and Bluenose-East Barren-Ground Caribou Herds Management Plan*. Terriplan Consultants (Ed.). Yellowknife: Department of Environment and Natural Resources, Government of the Northwest Territories. http://www.grrb.nt.ca/pdf/wildlife/caribou/CB_BNW_BNE_Mgmt_Plan_FINAL.pdf.

Agrawal, A., & Gibson, C.C. (1999). Enchantment and disenchantment: The role of community in natural resource conservation. *World Development, 27*(4), 629–649.

Ashley, B. (2002). *Edible Weights of Wildlife Species Used for Country Food in the North-west Territories and Nunavut.* Yellowknife: Wildlife and Fisheries Division, Department of Resources, Wildlife, and Economic Development, Government of the Northwest Territories.

Berkes, F. (1983). Quantifying the harvest of Native subsistence fisheries. In R. Wein, R.R. Riewe, & L.R. Methven (Eds.), *Resources and Dynamics of the Boreal Zone* (pp. 346–363). Ottawa: Association of Canadian Universities for Northern Studies.

Berkes, F. (1990). Native subsistence fisheries: A synthesis of harvest studies in Canada. *Arctic, 43*(1), 35–42.

Berkes, F. (2008). *Sacred Ecology: Traditional Ecological Knowledge and Resource Management* (2nd ed.). New York: Routledge.

Bluenose Caribou Management Plan Working Group (BCMPWG) (2011). *Taking Care of Caribou: Cape Bathurst, Bluenose-West, and Bluenose-East Barren-Ground Caribou Herds Management Plan.* Terriplan Consultants (Ed.). Yellowknife: Department of Environment and Natural Resources, Government of the Northwest Territories.

Bousman, C.B. (1993). Hunter-gatherer adaptations, economic risk and tool design. *Lithic Technology, 18*(1–2), 59–86.

Chiu, A., Goddard, E., & Parlee, B. (2016). Caribou consumption in northern Canadian communities. *Journal of Toxicology and Environmental Health. Part A., 79*(16–17), 762–797.

CircumArctic Rangifer Monitoring and Assessment Network (CARMA) (n.d.). "Bluenose West." https://carma.caff.is/herds/538-carma/herds/587-bluenose-west.

Davison, T.M., Callaghan, K., Popko, R., & Milakovic, B. (2009) *Population Estimates of Tuktoyaktuk Peninsula, Cape Bathurst and Bluenose-West Barren-Ground Caribou Herds Using Post-calving Photography.* Inuvik: Gwich'in Renewable Resources Board and Government of the Northwest Territories. http://www.enr.gov.nt.ca/sites/enr/files/239_manuscript.pdf.

Environment and Natural Resources (n.d.). Barren-ground caribou: Northern herds. http://www.enr.gov.nt.ca/en/services/barren-ground-caribou/northern-herds.

Environment and Natural Resources. (2011). *Caribou Forever – Our Heritage, Our Responsibility: A Barren-Ground Caribou Management Strategy for the Northwest Territories, 2011–2015.* Yellowknife: Department of Environment and Natural Resources, Government of the Northwest Territories.

Filion, F. (1980). Human surveys in wildlife management. In S. Schemnitz (Ed.), *Wildlife Management Techniques Manual* (pp. 441–453). Washington, DC: Wildlife Society.

Gilchrist, G., Mallory, M., & Merkel, F. (2005). Can local ecological knowledge contribute to wildlife management? Case studies of migratory birds. *Ecology and Society, 10*(1), 20.

Gunn, A., Johnson, C., Nishi, J., Daniel, C., Russell, D.E., Carlson, M., & Adamczewski, J. (2011). Understanding the cumulative effects of human activities on barren-ground caribou. In P.R. Krausman & L.K. Harris (Eds.), *Cumulative Effects in Wildlife Management: Impact Mitigation* (pp. 113–133). Boca Raton, FL: CRC Press.

Gwich'in Renewable Resources Board (GRRB). (1997). *Nành' Kak Geenjit Gwich'in Ginjik (Gwich'in Words about the Land).* Inuvik: Gwich'in Renewable Resources Board.

Haener, M.D., Dosman, W.L., & Boxall, P. (2001). Can stated preference methods be used to value attributes of subsistence hunting by Aboriginal peoples? A case study in northern Saskatchewan. *American Journal of Agricultural Economics, 83*(5), 1334–1340.

Indigeneous and Northern Affairs Canada (INAC) (1992). Gwich'in Comprehensive Land Claim Agreement. https://www.aadnc-aandc.gc.ca/eng/1427294051464/1427294299170.

Inuvialuit Joint Secretariat (2003). *Inuvialuit Harvest Study: Data and Methods Report 1988–1997.* Inuvik: Inuvialuit Joint Secretariat.

Keane, A., Jones, J.P.G., & Milner-Gulland, E.J. (2011). Encounter data in resource management and ecology: Pitfalls and possibilities. *Journal of Applied Ecology, 48*(5), 1164–1173. https://doi.org/10.1111/j.1365-2664.2011.02034.x

Kruse, J., White, R., Epstein, H., Archie, B., Berman, M., Braund, S., . . . Young, O.R. (2004). Modeling sustainability of Arctic communities: An interdisciplinary collaboration of researchers and local knowledge holders. *Ecosystems (New York, NY), 7*(8), 815–828. https://doi.org/10.1007/s10021-004-0008-z

Kuhnlein, H.V., Goodman, L., Receveur, O., Spigelski, D., Duran, N., Harrison, G.G., & Erasmus, B. (2013). Gwich'in traditional food and health in Tetlit Zheh, Northwest Territories, Canada: Phase II. In H.V. Kuhnlein, B. Erasmus, D. Spigelski, & B. Burlingame (Eds.), *Indigenous Peoples' Food Systems and Well-Being: Interventions and Policies for Healthy Communities* (pp. 101–120). Rome: Centre for Indigenous Peoples' Nutrition and Environment, Food and Agriculture Organization of the United Nations. http://www.fao.org/docrep/018/i3144e/i3144e.pdf

Kuhnlein, H.V., & Receveur, O. (2007). Local cultural animal food contributes high levels of nutrients for Arctic Canadian Indigenous adults and children. *Journal of Nutrition, 137*(4), 1110–1114.

McDonald, I., & Gwich'in Renewable Resources Board (GRRB). (2009). *Gwich'in Harvest Study: Final Report.* Inuvik: Gwich'in Renwable Resources Board.

McMillan, R. (2011). *Resilience to ecological change: Contemporary harvesting and food-sharing dynamics in the K'asho Got'ine community of Fort Good Hope, Northwest Territories* (MSc thesis). University of Alberta, Edmonton.

Moller, H., Berkes, F., Lyver, P.O., & Kislalioglu, M. (2004). Combining science and traditional ecological knowledge: Monitoring populations for co-management. *Ecology and Society, 9*(3), 2.

Moran, E.F. (1982). *Human Adaptability.* New York: Westview.

Natcher, D.C., & Hickey, C.G. (2002). Putting the community back into community-based resource management: A criteria and indicators approach to sustainability. *Human Organization, 61*(4), 350–363.

Nuttall, M., Berkes, F., Forbes, B., Kofinas, G., Vlassova, T., & Wenzel, G. (2005). Hunting, herding, fishing, and gathering: Indigenous peoples and renewable resource use in the Arctic. In C. Symon, L. Arris, & B. Heal (Eds.), *Arctic Climate Impact Assessment* (pp. 649–690). New York: Cambridge University Press.

Poppel, B., Kruse, J., Duhaime, G., Abryutina, L., & Marg, D.C. (2007). *Survey of Living Conditions in the Arctic: Results.* Anchorage, AK: Institute of Social and Economic Research, University of Alaska. http://www.iser.uaa.alaska.edu/Projects/living_conditions/images/SLICA_Overview_press.pdf

Porcupine Caribou Management Board (n.d.). About the herd. http://www.pcmb.ca/herd.

Receveur, O., Boulay, M., & Kuhnlein, H.V. (1997). Decreasing traditional food use affects diet quality for adult Dene/Métis in 16 communities of the Canadian Northwest Territories. *Journal of Nutrition, 127*(11), 2179–2186.

Sandlos, J. (2004). *Northern wildlife, northern people: Native hunters and wildlife conservation in the Northwest Territories* (PhD diss.). York University, Toronto.

Usher, P.J. (2000). *Standard Edible Weights of Harvested Species in the Inuvialuit Settlement Region*. Inuvik: Wildlife Management Advisory Council.

Usher, P.J. (2002). Inuvialuit use of the Beaufort Sea and its resources, 1960–2000. *Arctic*, *55*(5), 18–28.

Usher, P.J., Duhaime, G., & Searles, E. (2003). The household as an economic unit in Arctic Aboriginal communities, and its measurement by means of a comprehensive survey. *Social Indicators Research*, *61*(2), 175–202. https://doi.org/10.102 3/a:1021344707027

Usher, P.J., & Wenzel, G. (1987). Native harvest surveys and statistics: A critique of their construction and use. *Arctic*, *40*(2), 140–160.

Winterhalder, B. (1981). Optimal foraging strategies and hunter-gatherer research in anthropology: Theories and models. In B. Winterhalder & E.A. Smith (Eds.), *Hunter-Gatherer Foraging Strategies: Ethnographic and Archaeological Analyses* (pp. 13–35). Chicago: University of Chicago Press.

Part 2
Understanding Caribou

5
We Are the People of the Caribou

Morris Neyelle

The Sahtú Gotine – the people of Délı̨nę – have lived in the Great Bear Lake region for thousands of years. In addition to being an important caribou-hunting area, the area is a very important fishing area. It was traditionally a fish camp where people met each year during the summer months. The name "Sahtu" is thought to have originated from "sah" (bear) and "tu" (lake), meaning "Bear Lake."

There are many rich oral histories that have been passed on from our ancestors about the Sahtú region. The late elder George Blondin told and retold oral histories, including the story of the caribou boy, which tells how Dene people of the region were born from the caribou. As a result, we have a very close spiritual connection to caribou.

This photo of Leon Modeste (Figure 5.1) was taken at a fire feeding ceremony in Délı̨nę in 1995. The photo is important because it tells a story of the continued importance of elders in our community's way of life. The picture was taken at a spiritual place near Délı̨nę. It was an event attended by many community members, including youth who are carrying on our community's traditions, Feeding of the Fire.

The drum he is holding (Figure 5.2) is made out of caribou hide and birch-wood. It was made by another elder, Alfred Taniton. Finding the right kind of birch is important, and it can be difficult to cut and steam the wood to make a good frame. Not everyone has the skills to make drums. Once the frame is made, caribou hide that has been stretched thin and soaked for several days is fixed around the frame. Once the hide is attached with

FIGURE 5.1   Leon Modeste at a culture camp in Délı̨nę | *Photo by Morris Neyelle*

sinew (kind of like our traditional thread), it has to be dried carefully, or it will crack. We also stretch two thin, twisted caribou hide pieces across the drum once it dries. It creates a buzz or rattle when people play the drum. We usually use a stick handle made out of birch to play the drum.

Although Leon Modeste is pictured by himself, people usually get together to drum as a group. Men and boys drum and sing together at special gatherings. Before meetings of our chief and council, for example, drumming is done as a prayer to encourage people to work together under the guidance of the Creator. Drumming is also done for fun – everyone from the community gets together and dances the drum dance to cele-brate. There are some really good drum dances such as the Mountain Dene drum dance. The drum dance pictured in Figure 5.3 took place during a

FIGURE 5.2 Caribou hide and birchwood drum | *Photo by Morris Neyelle*

FIGURE 5.3 Drum dance at the New Years' celebration in Délı̨nę, 2008 | *Photo by Morris Neyelle* 2008

special gathering in 2008 to celebrate the New Year. The drum helps us connect with the Creator and helps us heal during hard times. There are songs for pretty well everything we do.

Caribou hunting continues to be an important part of our culture. Caribou is our traditional food, so we take care of the caribou. We hunt

FIGURE 5.4   Caribou hunters from Déline travelling on Great Bear Lake | *Photo by Morris Neyelle*

only what we need. Hunting today is done much like it was in my grand-father's time. People work together and hunt in the same areas with the same respect for caribou that we have always had. But technology has changed. Today, we use snowmobiles instead of dogsleds. This makes travelling and hunting expensive for many people since fuel costs a lot in our community.

Figure 5.4 shows a group of hunters on their way to hunt caribou. They have stopped to rest and drink tea on Great Bear Lake, near Caribou Point. This is an important caribou-hunting trail for us.

Caribou can been found in many areas during fall and winter. But sometimes our hunters don't come back with caribou; instead, they have harvested a moose or other food. We have lots of different kinds of foods in our area from the land. We also harvest a lot of fish. So if there is no caribou, that is part of the cycle. We know the caribou will come back some day, if people respect them and take care of the land.

6

Harvesting in Dene Territory: The Connection of Ɂepę́ (Caribou) to the Culture and Identity of the Shúhtagot'ı̨nę

Leon Andrew

For the Shúhtagot'ı̨nę (Mountain Dene), Ɂepę́ (caribou) is our main source of food and clothing. In traditional times, the Shúhtagot'ı̨nę were a very mobile people, always moving camp from one location to the next while searching for food. Today, we live in Tulit'a ("Where the waters flow together"), located on Dehcho ("Big river," Mackenzie River), just below the Arctic Circle – although we still travel into the mountains, our traditional lands. The Shúhtagot'ı̨nę lived from one generation to the next by following the rules that were laid out by our elders. For generations, our ancestors lived on the land, camping and searching for food daily. They lived in small groups of families, hunting to prepare enough dry meat for winter. Like them, we strongly believe in traditional medicine from the land and in the dream world.

The Shúhtagot'ı̨nę are known for raising their children to follow strict rules in camp and out in the bush. Youth are taught how to be a good person, not to touch other people's belongings, and to always tell the truth. The elders always said that to be a good person, you must use your ears and eyes in order to gather the best advice from your fellow people. These rules also extend to diet and food. For example, youth are taught to eat only healthy bush food that will help them to be fast runners, allowing them to chase down Ɂepę́ or moose in deep snow during the winter. Yamǫ́zha was an important culture-hero who made most of the rules we live by today.

As an aspect of the Shúhtagot'ı̨nę way of life, we have always taught our youth to search for and be respectful of supernatural powers. They achieved this in the past by camping by themselves for a length of time. Young men were taught to camp a fair distance away from the main camp, and only their fathers were allowed to visit them. In this quiet place, it was hoped that they would gain supernatural powers through their dreams. Young men were the future lifeline of Shúhtagot'įnę survival, and they needed to obtain įk'ǫ (supernatural powers) through dreaming. If a young man was fortunate enough to receive these powers in his dreams, he might become a mįdzita (Caribou master) or a great healer. This method was used for centuries by our ancestors. By living alone and searching for powers in the dream world while they slept, only a gifted few gained such powers.

There are four different kinds of spiritual practices that the Shúhtagot'įnę have maintained throughout the generations. Each of these practices is important and serves a distinct purpose that needs to be properly understood by our young people. We are aware that other communities have their own practices, and we respect that.

The first practice relates to the search for įk'ǫ. Historic sites like Nááts'įhch'oh ("Points upward like a porcupine quill," Mount Wilson) and Pəyɔ́káeké ("Mirage that appears in the distance," Mirage Mountain) are important dreaming places and help to provide įk'ǫ for the Shúhtagot'įnę. People who really want to gain įk'ǫ have to work at it and sacrifice their time, meaning that they must spend their whole youth living in the bush and periodically living alone and away from their parents. Even then, only a few special young people are awarded a dream gift. Our oral history tells us that many youth spent so much time in their dream camp that their hair turned white, yet nothing happened and they went home empty-handed.

The second practice involves drumming, singing, and drum dancing. These forms of expression have their origins in Néwhehtsįnę (the Creator) rather than in the earthly realm of įk'ǫ. They have remained very spiritual and important to maintaining Shúhtagot'įnę relationships with yak'ɔot'įnę (people of the spirit place). We see drum songs as a spiritual gift from Néwhehtsįnę that allows us to express our gratitude and respect for what we have been blessed with. Once the drummers start into a dance song, it becomes like an act of magic for them, and the drum song comes to them naturally. The drum songs tell us that they were given to Shúhtagot'įnę by the angels of the yak'ɔ (spirit place).

Feeding the fire is the third practice. Shúhtagot'įnę believe that the spirits of some ancestors who have passed on remain present on earth. By feeding the fire, we communicate with those spirits and ask them to help us with healing and good hunting.

The fourth practice is sharing gifts with the land. When we come upon certain spiritual places, we leave gifts for the land in order to maintain the cycle of sharing. The land gives us gifts, and we always give back. If we keep this practice alive, the land will respect us and the weather will remain calm.

Shúhtagot'įnę once used dog teams for transportation, driving the teams in single file. In summertime, they used dogs for packing and moving supplies. Shúhtagot'įnę used dog packs when they hauled tons of dry meat and fresh meat to the shore of the Begádeé ("Gravel river," Keele River), where they made moose-skin boats. They carefully navigated their moose-skin boats on the dangerous, fast-flowing river to safely bring their families and supplies out of the mountains to Tulit'a.

The Shúhtagot'įnę have always been culturally connected to Ɂepę́, hunting them from one generation to the next. In this way, they have made a living by hunting, maintaining their way of life from day to day in order to survive. In the old days, people hunted Ɂepę́ with bow and arrow and spears. They also used snares. Fences made of wood were set to trap the Ɂepę́. Snares were set in openings in the fences, and in that way, the herd was captured. Once Ɂepę́ were harvested, people would take the meat home and hold a feast to celebrate the gift of the Ɂepę́ harvest.

All parts of the Ɂepę́ were important, being used for many things besides food, including clothing, drum skins, lodge coverings, and dog packs. Babiche, made from the fresh hides, was braided into snares and strong rope. Ɂepę́ bones were smashed into small pieces and boiled till the grease was released. The bone fat would rise to the top of the broth and be collected for food. Ɂepę́ stomachs were used for collecting all the bone fat, making sure nothing was wasted. Stomach bags were also used for collecting blood, which was frozen and used later for soup.

The Shúhtagot'įnę have great respect for the Ɂepę́, and we take good care of them. Shúhtagot'įnę also believe Ɂepę́ are special animals that have the ability to travel long distances without tiring during their migration. People should not tamper with their migration routes. In the old days, Ɂepę́ fences and snaring sites were kept clean, and only the hunters were allowed to walk over the area. Salt licks are special places to the Shúhtagot'įnę. We would never butcher a Ɂepę́ on a salt lick and instead would drag the carcass away to dress it out. Youth are taught never to hit a Ɂepę́ with a stick or club, especially on the head, for doing so is very disrespectful.

Ɂepę́ bones must be discarded neatly in tree branches. Elders never left the bones unattended, and no one threw them in the water. It is especially important not to dispose of the bones in an open fire. Ɂepę́ skulls are sacred to the Shúhtagot'ınę, and elders always took special care of them, making sure they were placed in trees.

Our language, used by our people for hundreds of generations, is very descriptive and suited to the environment we live in. With our language, it is easy to describe the geographic features of our region, the lay of the land, and the way water sits and flows, to differentiate the many plants, and to describe the weather, stars, fish, mammals, and birds. There are many names to describe differences of gender and age among Ɂepę́. For example, male Ɂepę́ are called *mıhcho,* young bulls are called *yárego,* females are called *mıdzıh,* a calf is called *Ɂezhah,* and a cow and calf together are called *mıdzı dezhá.* We hunt the mountain woodland Ɂepę́ which we call *shúhta goɁepę́.* In the old days, sometimes we would see a Ɂepę́ from the north, a migratory Ɂepę́, which we called *tenatɫea* (long-distance runners). The elders say that tenatɫea come from the ocean shore. Migration routes are very important to the Shúhtagot'ınę. We call them *nı́ǫnep'ęnę.* They are like a directional system that guides the animals. A river channel, like the one the Begádeé flows in, is called a *tunıp'ęnę.*

The elders' role in teaching our youth is very important. Over the years, elders have guided young hunters from one generation to the next in how to respect animals like the Ɂepę́. They have also been taught how to take care of the animals' habitat. While parents were out finding food and doing other tasks – hunting, snaring, tanning hides – the elders were looking after the children. The youth learned the oral tradition by spending time with the elders and listening to their stories and teachings. The stories of Yamǫ́zha are particularly important because they teach us how to respect the animals.

Our traditional knowledge has taught young people about how the Shúhtagot'ınę survive and maintain their way of life. Traditional knowledge also teaches our young people about the wildlife and their habits. By practising our culture through the hunting of Ɂepę́, sheep, moose, waterfowl, ground squirrels, and grizzly bears, the young people come to know the habits of each animal.

Elders also teach the young men how to make hand tools like awls from Ɂepę́ dew claws, which can be used to punch holes in wood for snowshoes, or how to make beaver-teeth knives for carving out a hole to fit a wood bracket in place. They teach the young men how to make bows, arrows, snowshoes, willow-bark fishnets, spears, and many other tools.

Elders teach young people how to travel safely in all four seasons. They teach them how to travel over the ice during freeze-up or spring thaw and about water safety when travelling by boat. They teach them how to read the weather and how to travel in different terrain.

Elders teach young people how to select and cut down a tree for fire-wood. By actually showing them, they assure the youth's safety. Young people are taught how to read a tree before approaching it to cut it down. Most trees have weak tops, and when a person hits the tree with an axe at the base, the top of the tree can break off and come down heavily on one's head, knocking a person out and thus causing a serious problem, especially if one is alone in the wilderness.

The oral history and traditional knowledge imparted by the Shúhtagot'įnę Dene elders teach us what to do when ʔepę́ are scarce. In these times, we are taught to track back to the fish lakes in the front ranges and Dehcho lowlands, places like Tł'ok'átęnęʔa Túé ("Thicket of red moose willows that extends to the lakeshore," Tate Lake), Talǫ́ǫ́ Túé ("Open water," a lake that is very slow to freeze, Stewart Lake), Tets'ehxe Túé ("Hitting the surface with a stick to chase fish that migrate up the shallow creeks from the lake," Drum Lake), and Dehdeləjǫ Túé ("Good habitat for a sucker fish," Sucker Lake or Three Day Lake). We are taught about locations where there are lots of rabbits or birds. Although the Shúhtagot'įnę can survive only on rabbit for a long time, they say that doing so is not good for you. In the old days, the Shúhtagot'įnę used only willow-bark fishnets when fishing through the ice, and they had only stone axes to chop the holes. In this life, the young men needed to gain power through their dreams, as that was the only way they could survive.

Critical areas for the Shúhtagot'įnę included Tets'ehxe Túé, Įts'éo Pįę́ ("Open spot on the high hills that is good habitat for big bull moose," Moose Prairie Mountain), Įts'ét'o (Moose Nest Mountain), Tłįdedele Daıtł'ǫ (Red Dog Mountain), Poghócho ("Big valley," Summit Creek), and the Nezheįlį ("Water flowing through the ridge," Underground Creek). People helped one another by hunting in groups. Some would chase moose out of a mountain valley together while others sat at the mouth of the valley waiting for them to come. These techniques were necessary when we hunted with bows and arrows. People would catch small game like rabbit and ptarmigan using snares. People also helped each other with fishing by making willow-bark fishnets together. Sometimes they would even make a beaver net out of willow bark.

Shúhtagot'ı̨nę always helped each other. They would help families without children by getting firewood and snow water for them. And couples without children would return the favour by helping families with lots of children when it came time to move camp. In this way, the Shúhtagot'ı̨nę lived in harmony with each other and with the animals and land.

7

Dene Youth Perspectives: Learning Skills on the Land

Roger McMillan

A child, if he caught a rabbit, [his] first rabbit, his mother and father would make a feast out of that rabbit. They cook it, and they invite all the elders. With a first moose or caribou, they do the same. When they kill their first ... then they cook it for the elders. And it is a thankful thing for the peoples, for the ... child grows up ... proud of his first killing. That's ... how we lived.

— Elder Thomas Manuel

So when I first snared my rabbit, my old man ... and my grandmother ... those two ... they came, they started dancing ... the first time I snared my rabbit. Way down past Colville Lake, that's where we were living, when I first snared my rabbit, so the old peoples celebrated ... all the old peoples. So they cooked it and they made a big feast, and the old peoples used to say ... when they left, they said, "Thank you, thank you, because you're going to reach the age like my backbone is ... I ate your food well ... [and] you will follow me in my tracks to be an old person."

— Elder Gabe Kochon

In a culture of hunting, a young person's first kill is a momentous occasion, and each "first" – whether a rabbit, caribou, or moose – is remembered with reverence. Elders in Fort Good Hope also recall how these occasions were publicly celebrated in honour of a newly productive hunter who might add to the group's food supplies in times of need. During an interview over tea, as another elder recounted a similar celebratory practice, a youth in the adjoining room interjected, "Where was my feast?" He had recently shot his first moose and valued the occasion tremendously, but it had not been celebrated. Young hunters in Fort Good Hope certainly take pride in the same rights of passage that elders reminisce about, and they are always keen to describe their hunting exploits. But at the same time, their descriptions of how these events are valued by others are substantially muted in comparison with the fond memories of the elders.

Over the course of the past few generations, Western society's economic and political expansion into the North has promulgated the wage economy, created sedentary living conditions, and diminished the quality of the ecosystem (Nuttall et al., 2005). On my first trip to Fort Good Hope, it became evident that in the context of people's efforts to balance aspects of traditional and Western modes of living, the younger generations are viewed with particular concern. There are sentiments that knowledge and skills essential to the Dene way of life are less prevalent among the youth, the implication being that the community will suffer if traditional ways are forgotten. This chapter's focus is therefore on the opportunities for youth to learn and practise the traditional skills used in procuring Dene foods against the ecological backdrop of changes in the availability of barren-ground caribou. I conclude that although many youth between the ages of eighteen and thirty-five do not report having significant experience on the land, many also continue to seek expertise in traditional pursuits. Their interests are facilitated by learning opportunities such as community hunts and by practice opportunities such as hunting barren-ground caribou around Colville Lake. Furthermore, youth who are engaged in harvesting make significant contributions of Dene foods to other households in the community – especially to elders' households – through sharing networks.

SETTING

Fort Good Hope is a K'asho Got'ı̨nę Dene settlement of 600 to 700 people on the banks of the Dehcho (Mackenzie River), just below the Arctic Circle in Canada's Northwest Territories. It is seasonally accessible either by riverboat or by winter road and accessible all year by aircraft. Like in other northern communities, the social fabric of Fort Good Hope is nested in an assemblage of families who once resided over a wider landscape. Authors note between three and seven broad kinship groups based on geographical affiliations within the families that now reside in Fort Good Hope, including the lodge people (or Colville Lake people), the river people, and the mountain people (Hurlbert, 1962; see also Johnson &

Ruttan, 1993, p. 78; Kulchyski, 2005, p. 157; and Savishinsky & Hara, 1978, p. 322). Given the diversity of familial geographies, different subsistence priorities may continue to some extent as residents maintain affiliations with particular landscapes, such as moose habitat along the Mackenzie River and migration routes of Bluenose West barren-ground caribou around Colville Lake. But more worryingly, many in the community do not often experience any of these lands directly. Indeed, in the mixed economy now apparent, it is not uncommon to pursue a livelihood of regular wage work, although the opportunities to do so are limited. Younger people often leave the community to pursue higher-quality education in Fort Smith or Yellowknife or to seek employment in Norman Wells or elsewhere. Certainly, moving away from their home is a difficult and identity-forming process for young people, and through this experience Fort Good Hope comes to take on paradoxical meanings. Close social and familial bonds of identity provoke nostalgia quickly after one leaves, which is now augmented through extensive use of online social media by community members, but shortly after one returns, these bonds can become stifling or irksome. One participant in his mid-thirties describes coming back after a year away in Yellowknife:

> I always miss home, you know ... like even coming back from Norman Wells to here by skidoo, like I was gone for, I think ... almost just over a year – I was going to school. I was living in Yellowknife, and I was going to school there, so I couldn't just come back any time I wanted to, even though I had lots of chances to come back ... because I was going to school, or because I was working. And then, finally, Christmas came around, and I was going to school in Yellowknife and ... they said, "Oh ... there's a [charter] plane going to Norman Wells. You only have to pay fifty dollars. All the way right from Yellowknife to Norman Wells. And you only have to pay fifty bucks and that's it." I was like, "Oh yeah?" But I was like, "Then what?" And they said, "Then you're on your own from there to Good Hope." I said, "Okay, I'll go." So I gave them fifty bucks, jumped on a plane, and ... I started thinking, "I wonder if I should just ... maybe stay in Norman Wells for a couple of days before I go home." But then, I got there, and then I seen some of my friends, and then they told me, "Oh, we're going to be driving back by skidoo, come with us." I was like, "But I got no warm clothes." They said, "Well, borrow ones 'cause you can bring it back when you go back to Yellowknife." So I was like, "Yeah, okay." So I went to my friend's place and I said, "Okay, I need some ... skidoo boots, I need some warm mitts, I need a parka if you have it, ski pants ... and all

that [to] go back by skidoo." So they were all nice enough to just give it to me ... and then ... when we took off, like the closer we got to here, the more I felt at home, you know? ... Like, I don't know if you know where Apache Pass is ... You know when you're driving between those two ... that little valley there, right on the hill there? That's Apache Pass ... Well, when I was young, before they even had a winter road, that's where my trap line used to go. And that's probably the last time I ever trapped, I guess – ever since they made that winter road. That was probably the last time I ever trapped ... Yeah, and as soon as we started getting closer ... all these little places that I've been ... from around Apache Pass, back this way, even maybe a little bit towards the Wells. The closer we got back to here, the more at home I felt. And we even stopped, and my buddy told me, "Feels like home?" I told him, "Oh yeah – like really, man!" And he said, "Yeah." He said, "Been away for how long now?" I told him, "Over a year." And he said, "It feels good?" I told him, "Really, man." I told him, "I just can't wait now – I want to get home!" And ... so it was, it was good. I mean, just seeing all these places that I've been and that before. But yeah, I guess, you know, being away for a while and then coming back, it's ... it's a good thing. But then, once you get back here, then you start thinking, "What the fuck did I do that for?"

YOUTH INVOLVEMENT IN THE HARVESTING ECONOMY

Hurlbert (1962) remarks on youth leaving Fort Good Hope and describes their difficulties and pressures to return home, which reminds us that mobility for people in northern communities is not new – although it may be increasing. But mobility itself does not accurately reflect the actual opportunities available to northerners in larger centres. Salokangas (2009) gives a particularly good account of this incongruity as it relates to youth in Tuktoyaktuk and the meaning of education; she concludes that although youth in contemporary Tuktoyaktuk have ambitions and aspirations indicative of a southern life, they are very much constrained in terms of practical options in a small northern community. In terms of the characteristics of mobility, too, the alternatives are, by and large, other sedentary communities at varying psychological distances from their landscapes. It is this separation of people from the landscape that has been mourned by so many northerners and researchers documenting the settlement of dispersed and fluid bands into permanent communities in the 1950s and 1960s (Asch, 1979; Helm, 1965; Savishinsky, 1974). Possibly the gloomiest of these accounts is given by Savishinsky (1974, p. 219), who describes in terms of "stress" some of the consequences of the K'asho Got'ıne people's transition from seasonal bands to permanent communities, particularly lamenting "nascent urban settlements" such as Fort Good Hope. He contends that this settlement and the decline of the trapping economy diminished people's competence in living on the land, reduced personal effort and ability as a source of status and esteem, and disrupted the ethic of generosity because of a growing commitment to contradictory Western values (ibid., pp. 219–20). This image starkly contrasts with Hurlbert's (1962, p. 52) description of the pastimes of older Fort Good Hope boys, who "like to borrow canoes and kickers and go off for the day up and down the river visiting the various fish camps, shooting ducks and stopping to see how many fish are in each net. Younger boys will take a hunting canoe and go away from town to explore and hunt. The girls like to go on berrying picnics."

Although debates continue about whether change in the North should be characterized in terms of adaptation, assimilation, or development, the literature reflects a conception of Indigenous identities as resilient and inexorably connected to a sense of place on the landscape. Fajber (1996, p. 54) quotes a research participant who asserts, "Out on the land you know you're Dene. You know what to do. It's within you. Respect comes out. You are Dene and you are in touch with the land. Everyone has a role, everyone knows what to do." Place and purpose are therefore highly interconnected. Parlee et al. (2006) also portray such linkages in terms of country food harvesting, which they connect to conceptions of health, autonomy, language, and cultural longevity, or transmission, in their exploration of the Dene way of life. Hickey, Nelson, and Natcher (2005) have also referred to studies concluding that young Dene males are best able to cope with challenges they face in their communities if they maintain strong ties to the bush, and these authors themselves even describe the benefits of conducting interviews on the land rather than in town.

Nonetheless, in many accounts of northern communities, youth are referenced as examples of people's disconnect from the land and from traditional livelihoods (Ford et al., 2007; Hickey et al., 2005). In Inuit communities, Gombay (2009) records less consumption of traditional foods among younger people in Nunavik, and Kishigami (2000) expresses concern about a diminished ethic of generosity among the youth, forecasting negative implications for traditional food-sharing networks. Thus the position of youth is often regarded as one of separation from the land, with negative implications for the transmission of critical cultural skills and knowledge. The significance of knowledge – what has usually been called traditional or Dene knowledge – is thus a recurring theme in these accounts (for more on traditional knowledge, see Auld & Kershaw, 2005; Johnson, 1992; Nadasdy, 2005; Nelson, 2005; and Wray, 2011).

Condon et al. (1995) and Collings and Condon (1996) cite several factors in younger people's waning participation in country food harvesting in Ulukhaktok, including a lack of funds for equipment, changing dietary preferences, a lack of interest in economically marginal pursuits such as trapping, an increased dependence on wage employment, which limits free time, and an "addiction" to sports, which they contend to be a new mechanism for obtaining social status among young males. They also highlight the challenges facing their younger study participants as a result of being raised in sedentary conditions, noting that they had not received as much training in harvesting as the older group had gained living in smaller, more scattered seasonal settlements. The transmission of traditional knowledge is the focus of Ohmagari and Berkes (1997) in the context of the Cree Nation of James Bay. A primary conclusion of the authors is that the sedentarization of communities has disproportionately alienated women from land-based activities. They identify four social conditions that work against the transfer of traditional knowledge: a changing educational environment, insufficient time on the land, the development of bush skills at a later age, and changing values, which are linked especially to southern-style influences through television. Pearce et al. (2011) take this work further in their study community of Ulukhaktok, finding that younger participants (18–34) had typically not reached levels of competency equal to those of the older cohort (35–49) in the eighty-three specific land skills identified. They note the effects of residential school, lower fluency in Inuinnaqtun and Inuvialuktun, changes in family structure, and more distant caribou movements as factors inhibiting the transmission of such land skills. Meanwhile, in a Cree community of northern Alberta, Hickey et al. (2005, p. 292) document people's expectations that the next generation will experience declining access to bush resources due to shortages of time, finances, and knowledge.

But much has also been documented about programs designed to better acquaint younger people in Aboriginal communities with local ecological knowledge and bush skills. In most cases, these efforts are conceptualized holistically in terms of cultural preservation and revival. In Alaska, some schools have sought to integrate traditional ecological knowledge into classroom lessons about environmental issues (Barnhardt, 2005), and teachers themselves attend cultural-immersion camps on the land (Kawagley & Barnhardt, 1999). Arguments have been made, however, that the intrinsic experientiality of traditional knowledge defies integration into a classroom context (Barnhardt & Kawagley, 2005). In northern Alberta, Hickey et al. (2005) note elders' reluctance to utilize structured patterns of teaching and knowledge consistent with a classroom setting, as the knowledge and skills of import require an appropriate context for effective transmission. In an Inuit community, Bonny (2008) describes some frustration among youth at elders' unwillingness to respond to their inquiries, after encouraging youth to ask questions in the first place. Ford et al. (2007), Peloquin and Berkes (2009), and Pearce et al. (2011) all emphasize that participation in on-the-land activities is essential to the development and transmission of ecological knowledge, and they suggest that place-specific on-the-land camps can create favourable conditions for such participation to occur and can also strengthen interpersonal social ties of the participants. Ingold (2000) refers to this process of "enskillment" as involving a pedagogy of "directed attention" that precedes the lengthy period of practice required to gain proficiency in harvesting.

The interrelationships between culture and learning are outlined by Kumpulainen and Renshaw (2007, p. 110), who consider culture to be a "situated resource – a fund of knowledge and a repertoire of practice – that learners draw upon to make sense of their social and material world and to participate in it." The authors then characterize learning as a "process of enculturation and transformation into different community practices ... positing an agentic learner whose capabilities are afforded and constrained by the cultural tools they can *access* within their social setting" (ibid., p. 111, italics in original).

This study with K'asho Got'ı̨nę youth thus considers their own perspectives on Dene skills and knowledge, how they are pursuing them, and what opportunities facilitate their development. In particular, I consider these factors in relation to two hunting cases: a 2009 autumn community hunt and individual hunts for barren-ground caribou. My trajectory recalls Fajber's (1996) and Bonny's (2008) interest in countering popular discourses of problematization that may discount optimistic signs, such as the point made by Condon et al. (1995) that some young adults grow into hunting as other providers in their communities become elderly. Following Kumpulainen and Renshaw's (2007, p. 110) portrayal of the culture-learning interface, I acknowledge that "to investigate learning as an ethnographer ... is to focus on the practices and understandings of the members of a community, and the interactive processes that establish and maintain such practices."

METHODS

This study was conducted between August 2009 and September 2010 and was based on a variety of methods, including participant observation, semi-structured interviews with eighteen harvesters, a public workshop, and a youth focus group. I also lived in Fort Good Hope for five months in the process of conducting the research, trying to participate in community life as much as possible. This participation involved various activities, such as attending public and youth council meetings, hanging out in the Renewable Resources Council office, crawling through river islands to flush out moose, coaching biathlon for interested youth, and making uncoordinated attempts at square dancing.

Participant observation methods relate primarily to two hunting activities: the September 2009 community hunt and a series of individual hunts pursuing barren-ground caribou in November 2009. Interviews with all harvesters were conducted by the researcher seven to fourteen days after they returned from their hunts, with structured questions focusing on their perspectives on the most enjoyable parts of their trips and on how the resulting meat was distributed to others. These questions typically led to less structured conversations on topics ranging from the importance of hunting and sharing food to the social fabric of the community generally. A public workshop attended mostly by elders in September 2010 was an opportunity to gather feedback on some of the information compiled through the interview process and to gain an appreciation for the importance of hunting and sharing traditions to the survival of the K'asho Got'ɪnę people. Finally, a youth focus group arranged in November 2009 significantly crystalized young people's perspectives on traditional pursuits and has informed the premise of this chapter. It is critical to note that there are significant study limitations due to how the researcher's own cultural background has influenced the data collection and interpretations, as well as the possibility that participants may have omitted certain influential issues for various reasons (Bonny, 2008).

Categorizing the demographics of participants was not straightforward, but given the small sample size, only three groups were discerned on the basis of their age in 2009: youth born in 1974 or later (aged 34 or younger), adults born between 1950 and 1973 (aged 35–59), and elders born in 1949 or earlier (aged 60 or older). It is recognized that there may be substantial variation in appropriate definitions of "elder" (Davidson-Hunt & Berkes, 2003). Regarding the two hunting activities – the community hunt and the individual caribou hunt – only one of the harvesters is present in both samples.

It was difficult to get to know people in the younger demographic in Fort Good Hope, as their circumstances with family, friends, housing, employment, and school often seemed more complex than the circumstances of the older cohorts. Many in their twenties had young families of their own and sometimes struggled with the responsibilities involved. Job opportunities and housing were perpetual issues. For many people of all ages, pursuing formal Western education seemed to be a greater priority than focusing on traditional skills. The difficulties young people faced in earning their community's respect for their traditional knowledge and skills were often palpable in comments made in public meetings. Disapproving or concerned remarks were also made about youth who had no taste for country food. In fact, a popular story describing why caribou have sometimes avoided the vicinity of Fort Good Hope in the past features a "young person hitting a caribou with a stick." Such stereotypes of improper harvesting techniques continue to be commonly attributed to youth, with some people even viewing less accessible caribou as a positive circumstance, as inexperienced young hunters will then not be able to harvest them inappropriately. Thus youth not considered proficient at traditional pursuits may encounter disapproval for trying to practise them.

Youth Focus Group (November 2009): How to Build Knowledge and Skill

I never shot an animal before ... but I have been with my dad when he shot ... My first time seeing him shoot a caribou or a moose ... [was] when I was ten. And then ... [it] felt good, 'cause just down ... just a few ... about a mile down, my grandpa shot one, too. So we're both ... I don't know. I felt good helping them cut it up. And ... bringing it back in, sharing it with my grandparents and ... family. I think eating brings everybody together.

Male and female youth attending the focus group ranged in age from eighteen to twenty-five, and only a few described participating frequently in activities on the land. As a result, their comments about harvesting and food sharing were more often framed as observations about community norms, as opposed to personal accounts. Thus the discussion turned more toward what opportunities exist for youth to go onto the land and how youth value these experiences. The dialogue below is presented in the form of a metanarrative to portray the main flow of the discussion. The facilitator began with a question about how youth might perceive the community's different expectations of them.

[There are] a lot of expectations to do this and to do that. And there's only so much we can do. And ... the elders keep on saying, "You have to go and learn how to live off the land." And your parents say, "You have to go to school, get your education." And it's so complicated and diverse these days. And ... it's not our fault ... I don't know, like sometimes the way they put it ... it kinda sounds ...

They doubt us ...

Some feel sorry for us because of the way life is these days ...

That we're ... this old, and that we don't know ... yeah.

... Anybody older, who knows they're ... better than you at it ...

Yeah, more people over forty would feel sorry for us or ... just ... yeah, judge.

In terms of gaining the trust of older generations, youth emphasized the need to prove themselves. This was evidently considered difficult, requiring demonstrations of substantial skill and experience, before which they would not be trusted to go into the bush. Participants were clear that the process of building trust was a lengthy one:

Maybe they have to go out more often?

They gotta pay their dues ...

Gotta know how to survive, how to skin ... kill ... be ambitious ... not lazy.

You gotta be independent, responsible, reliable, up and ready to go anywhere.

Yeah. Takes years and years and years and years. Can't just say, "Oh, hey, I'm going to grab my gun and I'm going to go." They'll ask you where you're going to go, ask you who you're going to go with. Are you dressed up warmly? Lots of questions if you barely go. They'll make sure you got everything. Like you're a greenhorn at first ... Then after you show them you're not a greenhorn, then, yeah ... it's alright.

Yeah, 'cause you can't ... you can't just go and show them.

The conversation thus turned toward the ways that youth begin this long process of becoming familiar with the land and toward the traditional skills needed to make a livelihood, at least in part, from hunting, fishing, and trapping. Youth in the focus group emphasized the role of their families in introducing them, or not, to pursuits on the land:

It kind of depends on where your family came from. 'Cause, like, comparing our families, like my family doesn't really go on the land, like at all. And hers does. So it kind of depends on where you come from ...
Sometimes ... I didn't have a choice when I went on the land. I didn't want to go ... but then at the end of the trip ... I'm ... I'm glad I went. And ... it was an experience and it made me get closer with my family. But now that I notice that we don't ... go out a lot. We don't have that family bond as much ... but ... you do get a lot from it. My dad, he was brought up by his mom. So then ... my granny liked to drink a lot. So they didn't really get to go much ... out on the land. I think that's what affects most of ... most of us. Like, think some of our parents drank ... drank a lot ... then it did affect us, either way. But if you grew up with your grandparents, then if they didn't drink ... then they'd probably show you a better way. But ... having parents that drink was ... probably would be ... would ... yeah, affect the time you go out.
Me, too, I never even learned a thing from my dad about the bush. It was always someone else that taught me ... I'd say if I stayed here [when] I was like a little kid, I'd go out on the land more, but ... um, like when I was younger, my mom was going to school, eh, so I never came back here until I was eight. But then, yeah, came back here and I learned. But then that was one of the challenges she faced – to getting me back here and getting me to learn my culture.

The focus group participants also explained the difficulties that families faced in balancing work, the costs of living, and the costs involved in going onto the land. These difficulties were seen as increasing, and the requirements for employment were seen as becoming more stringent. In general, opportunities to go into the bush were considered more constrained:

> It's harder for parents to live ...
> Yeah, or you need your Grade 12 to get a good job. Back then it was just ... you work, you work. Any job was good. There was more jobs back then, too. So it was easier to live that kind of life. But now it's getting harder ...
> I don't know ... I'm thinking maybe it's harder to go out on the land now 'cause, like, people that want to teach us are, like ... they're too busy or something. They can't, like, offer bringing you out. And ... "Let's go on the land" ... They can't, like, just ... do it for free ...
> It's kind of hard to take youth out, too, 'cause mothers or fathers they don't ... You gotta establish the trust first. Before you take their kids out on the land ... make sure they're safe.

LEARNING TRADITIONAL SKILLS: A COMMUNITY HUNT

One theme in the focus group dialogue above addresses the difficulties that families face in inculcating traditional skills in those growing up. Participants indicated that occasions for youth to learn traditional skills from knowledgeable active harvesters and elders are sporadic. There are some concerted efforts to counter this problem, but school programs incorporating traditional knowledge are inconsistent, and on-the-land youth camps are of limited capacity. Hickey et al. (2005) record similar conditions in First Nations communities of northern Alberta, as do Ford et al. (2007) in Inuit contexts. Nevertheless, teaching traditional skills to youth is a priority for many hunters and for the Fort Good Hope Renewable Resource Council (RRC), a local agency tasked with bolstering hunting and trapping livelihoods. On hunts subsidized through the RRC, the attendance of both youth and elders is a primary concern. An emphasis on intergenerational learning through community hunts is also described by other researchers working with different communities (Brook, 2010; Pearce et al., 2011) and is a significant priority for funding agencies, such as the Brighter Futures Program (GNWT, 2000, p. 55). After the 2009 autumn community hunt, the hunt leader described his motivations to maintain connections between local youth and the landscape:

> I know it's very important to use the land because that was the old teaching. By being on the land, you keep the land alive and the land keeps you

alive. It's like an exchange. And so ... my grandfather's side, we are the Mountain people, and so you always talked about the mountains. Now, they enter land claims, self-government, I'm saying, like ... we should leave kind of marks on the land, where we've been. And so our kids, that way they can use those places ... My grandchild with us, he knows more about that [area] surrounding where we were, Tabasco Lake area. It's true – that area, that's where the trail ran to Mayo, and so it's important ... He's just beginning to understand that land a little bit more.

In late August or early September, as Subarctic colours begin to change, as the insects subside, and as the air becomes crisp, Dene communities across the Northwest Territories organize expeditions to harvest barren-ground caribou migrating into the treeline. The animals are in prime shape, their hides thick but not yet spoiled by fly larvae. These collective hunting forays have a long history. Helm (1965) notes crews setting off for caribou mid-century by canoe, and Hall (1989) describes traditional hunting "corrals" and rock piles built to direct a caribou herd toward a waiting group of hunters. Fort Good Hopers themselves also talked of their ancestors walking from the Mackenzie River out to Horton Lake, 300 kilometres away, to intercept the migrating herds. The 120 residents of Colville Lake have for many years essentially relocated to Horton Lake for two weeks every autumn by float plane, leaving only a few caretakers in their community. Although Fort Good Hopers had joined Colville Lake residents at Horton Lake in previous years, in 2008 and 2009 the Fort Good Hope Renewable Resources Council organized a separate community hunt in the Mackenzie Mountains, notably targeting moose, mountain caribou, and Dall sheep rather than barren-ground caribou. In addition to being less expensive in terms of travel costs, this hunt allowed organizers to bring elders and youth together on the land, where they were able to reconnect with an important traditional territory.

In September 2009, five youth, four elders, six adults, and three external researchers flew by Twin Otter to Tabasco Lake in the Mackenzie Mountains for a community hunt, building a bush camp and spending nine days hunting and processing meat. After the hunt, I arranged interviews with all the local participants. The personal satisfaction of long periods in the bush was the most widely reported benefit among them. The opportunity for intergenerational learning was also appreciated according to the youth, especially regarding traditional women's skills. Given Ohmagari and Berkes's (1997) thesis that the settling of communities more abruptly removed women from land-based livelihoods and knowledge, community hunts may represent a particular opportunity to reacquaint young women with those roles. Women were absolutely central to the community hunt endeavour, cutting up harvested game and drying the meat. Although the men helped out with these tasks as well, the harvest amounted to at least 900 kilograms of meat (extrapolating from Usher, 2000; and Alistair Veitch, pers. comm., 2010), all of which had to be carefully processed (see Chapter 8).

Q: So what was the best part about the community hunt?

A: The cutting up of meat and everything. The butchering of meat. Making dry meat. And fleshing the hide – things I've never really done before ...

Q: So what was the purpose of the community hunt for you, or why is it important?

A: To keep the ... culture alive, I guess. To keep our traditions alive, to keep knowing how to process meat ...

Q: What was it like getting back?

A: Different. I dunno – it made me want to be out there more. It made me want to be going back out on the land even more. And doing a little bit more of the ... the things that the women do. Like flesh the hide and make moose hide.

In these interviews and others, it was clear that the opportunity to learn was definitely appreciated by male youth as well, sometimes evidenced most poignantly by those who regretted *not* participating in the hunt itself. One young informant said, "Without that community hunt, like, probably no young ... young people would have been able to see that or learn off that ... like, how to hunt in the mountains. 'Cause it's different than ... hunting, like, for caribou ... It's different eh, you learn."

Participants also kept track of a suite of harvesting equipment that male hunters either owned or had available to them (i.e., a rifle, snowmobile, riverboat, and cabin). Access to equipment is a significant factor in the ability to harvest (Condon, Collings, & Wenzel, 1995; Pearce et al., 2011, p. 286), and hunters with more equipment are expected to harvest more often than hunters with less equipment. Interviewees sometimes found it difficult to articulate responses, as multiple people (e.g., a father and son) from a single household may not have had the same harvesting equipment available to them. Among the eight male participants in the community hunt, the three adults and one elder typically had more equipment available to them than did the four youth. Two of the adults in fact had all of the needed equipment, whereas one of the youth had none of it. This finding aligns with Condon et al.'s (1995) account of Ulukhaktok, which describes the significant length of time and effort involved in slowly assembling the expensive array of equipment necessary to be an effective harvester. The total cost of a full set of harvesting equipment was estimated at $13,439.00 in 1985 (Gombay, 2009, p. 121).

Thus the community hunt is a learning opportunity that gives younger, less active harvesters access to the expertise and equipment of more active ones. In a broader sense, community hunts may also be important opportunities for participation by those of other demographics who lack a means of hunting, such as elders who have much to teach but who are no longer physically able to travel long distances on the land.

Although collective hunts may be valuable opportunities for youth to explore new locations and learn directly from elders and adult harvesters, many youth also seek to harvest by their own means in order to provide for themselves and their families. After the first snowfall, barren-ground caribou are anticipated around Colville Lake, and communication networks surge as their whereabouts are traced. Youth and some adults check Facebook, whereas older hunters call family members in Colville Lake for updates, and everyone scrambles to prepare their snowmobiles for the first hunting trip of the winter season. Meanwhile, meat and fish have already begun to arrive in Fort Good Hope, with elders retrieving packages sent by their Colville Lake relatives on the scheduled air service. Normally, in December private contractors begin to push through a winter road 176 kilometres from Fort Good Hope to Colville Lake, but hunters on snowmobiles are typically able to make the trip much earlier. Although creeks are still open and a flat landscape poses challenges for navigation, the first person to arrive back with a sleighload of meat in November 2009 was a renowned young hunter who had spent freeze-up in Colville Lake hunting and trapping. He was proud of breaking the trail. After hearing this news, the researcher began to make arrangements for renting a snowmobile and joining a pair of adult harvesters in pursuit of caribou the following week. We were probably among the first ten snowmobiles on the road.

After two years of work relating to communities and barren-ground caribou, I was thrilled to finally see one of these animals emerge in the distance from the bushy shores of Colville Lake and make its way out onto the ice, followed by two more. We diverted up into the trees where I rolled my skidoo, again. GT noticed and came back to help. I hate skidoos. I had nearly pinned myself underneath the machine in an icy creek the previous day, arriving in Colville Lake at midnight with my beaver mitts and rubber boots all soaked and generally feeling like a liability. The five-hour trip had been dark and bumpy – with only a few inches of snow on the ground and the creeks not completely frozen. The following morning, I spent some significant time jiggling skidoo wires, adjusting the choke, changing spark plugs, and generally failing to start the machine, to be assisted eventually by ten-year-old JC. He seemed perplexed that anyone could *not* know all about skidoos and authoritatively removed the spark plugs, poured some gasoline in a metal pan, and lit them all on fire, while I attempted to contemplate the experience of being raised up in Colville Lake. Charred plugs seemed to do the trick, and I set off after the hunters, finding them some way up the eastern shore. We were met soon after by JC and his father (I was supposed to have followed them, I realized), who would guide us up to the north end of the lake.

These caribou we saw were half-way up the lake; they had also noticed us and started to run farther out on the ice. The hunters let them go and we continued to the northern shore. At the north end, a maze of fens, stubby trees, and frozen muskeg was pocked by caribou tracks and lichen-bottomed potholes where they had been foraging. The hunters wove a circuitous route through the frozen swamps, following tracks or other indications I could only imagine. Eventually, we emerged back onto the lake, and our guides left to go back to town. Shortly afterward, we saw another herd (of five) out on the ice running west chased by what looked like a wolf (later I was told that this was actually a young caribou trying to keep up). The hunters directed me to cut left to divert them toward the shore. I did, and the herd took off in the planned direction. Following them, I saw the herd dash up the bank above the lake; one turned around on the crest of the bank and emphatically reared its large antlers. It kept doing this, each time turning around in a small circle. ML and GT got two from the herd and butchered them up while I made a fire.

It was dark when we arrived back in Colville with a total of seven caribou (three in one sleigh, four in another). The last three the hunters had found in the twilight, and with aching backs, they had to cut them up in front of their skidoo headlights. We stored the sleighs full of meat in a shed, and we all went in for a good meal with our wonderful hosts – and some rest. (Compiled from field notes, November 8–10, 2009)

In interviews following the hunt, a list of equipment owned by or available to the seven caribou hunters was recorded. The caribou hunters tended to be more moderately equipped than community hunt participants; no one possessed all of the harvesting equipment on the list, but neither did anyone lack everything. Similar to community hunters, the three youth in the contingent were somewhat less equipped than the four adults.

Although the trip seemed like a significant endeavour to me, it may actually represent one of the simpler hunting options seasonally available to harvesters in Fort Good Hope. Snowmobiles are far more affordable to purchase and maintain than riverboats, and many more youth have access to snowmobiles than boats. One youth explained, "This summer I didn't even go out for moose ... couldn't do it 'cause I didn't have no boat or anything." Thus although going to Colville Lake is a long trip, it is likely a more practical undertaking. It is reasonably safe, there are comfortable accommodations upon arrival, and hunters are able to visit family and friends. In addition, compared with hunting moose on the Mackenzie River, which often requires several cooperating hunters, caribou are relatively straightforward to track, harvest, and butcher individually or in pairs. Young hunters or others with limited means therefore have an opportunity to learn for themselves through harvesting practice. This hypothesis is further supported by responses to interview questions concerning hunters' motivations for travelling the 170 kilometres each way to Colville Lake by snowmobile to hunt caribou, when moose or woodland caribou might be available at shorter distances (see Table 7.1). Hunters

TABLE 7.1   Motivations to hunt for caribou at Colville Lake

Hunter	Comment (paraphrased)
Adult	I just enjoyed the time that we had there; visited people in Colville Lake and at a camp at Aubrey Lake. Caribou each got $700 worth of meat, so I don't have to buy it.
Adult	Good scenery out that way ... new country, and the people.
Adult	Caribou are easier to get. It's good to get out of town and see other people and your relatives down there.
Adult	It's fun and a challenge to see new places out past Colville Lake. It's easy for people in Colville Lake to get caribou because they're so close.
Youth	Caribou meat is soft – tastes good! You can visit family down that way ... visit people.
Youth	I don't like going for moose on the river – you have to use other people's cabins, too much hassle, and most times you don't even see anything. I prefer caribou.
Youth[*]	Caribou always follow the same route, so you'll be sure to bring back meat. It's hard coming back with nothing. When it's time to get caribou, I get caribou ... plus the thrill of the hunt; it's more fun than moose. You don't have to drive for hours looking for it, and you're going to have all this meat to bring home; the family will be happy, and you can give meat to those who don't have the privilege to go.

[*] Accompanied a harvester to Colville Lake.

tended to respond that when caribou are around Colville Lake, they are very likely to make a successful harvest. In addition, hunter responses emphasized that it is good to be out hunting generally, that hunting toward Colville Lake provides an opportunity to visit with friends or relatives, and that the meat is of great value to themselves, their families, and the community.

Sharing the Meat

There are numerous accounts of the importance of sharing out harvested meats within northern communities, which tend to regard the practice as promoting equity and maintaining social cohesion (e.g., see Asch, 1979; Barnaby, 1976; Collings, Wenzel, & Condon, 1998; Gombay, 2009; Kishigami, 2004; Parlee, Berkes, & Teetl'it Gwich'in Renewable Resources Council, 2006; Savishinsky, 1974; and Wenzel, 1995). Caribou meat is also highly prized by many in Fort Good Hope. Both adult and youth hunters often described at length their favourite methods of cooking up caribou delicacies – especially the head, the brisket, and the organs. Many of these meals are shared with guests, and large quantities of meat are given out to others.

By far, the most widespread response to questions about food sharing emphasized the importance of distributing meat to those who are considered less able to access traditional meat for themselves; within this group, elders are consistently a priority. Accounts of sharing with those less able to harvest for themselves usually depicted the interactions as one-way transfers, although participants sometimes described how these occasions were particularly special for themselves as well:

> Yeah, and ... when I get back, I drop off meat with my granny. My family ... or whoever's asking for meat. Mostly, I like to go to elders ... Yeah, drop off meat, tell them in Slavey how much I shot. And where I shot them.
>
> *Interviewer: Oh, right on. They like hearing the stories?*
>
> Yeah, especially from a young guy telling them in Slavey ... [They're] pretty impressed.

Tracking some basic sharing patterns of caribou hunters reveals an interesting result: more occasions of sharing with elders were reported by youth than by adult harvesters (see Table 7.2). This study was unable to

TABLE 7.2 Caribou hunting and sharing characteristics, November 2009

Male hunters	Number	Caribou harvested	Sharing events	Sharing with elders
Elders (≥60)	0	0	0	0
Adults (36–59)	4	16	31	6/31
Youth (≤35)	3	12	26	9/26
Total	7	28	57	15/57

record the size of each transfer, and such a small sample is not intended for generalization. However, it does bring full circle this chapter's discussion of intergenerational connectivity, which began with the observation that young men proud of their initial harvest do not gain public recognition in the same ways recalled by their elders. In this example, young people were clearly contributors to their community, harvesting and sharing in proportions equal to those of the adults, while making sure to provide for those same elders.

* * *

In February 2011 I was introduced to the anonymous Facebook profile of "Kahsho Gotine," who worked "for food," attended school on the land, and was rapidly accumulating friends in Fort Good Hope and elsewhere. It would be inappropriate to characterize life on the land as a core interest of all young people in Fort Good Hope given that some are certainly content to pursue livelihoods and recreation within the community and other urban centres, but there remains a palpable connection between the land and the idea of being K'asho Got'ı̨nę. Thus, for many, knowledge of the land and familiarity with the skills associated with a livelihood on the land are high priorities. It is also clear that youth seek the esteem of their elders and often consider traditional pursuits such as hunting to be paths toward this goal. For elders, women, and prospective young hunters with otherwise insufficient means, the community hunt seems to provide an opportunity to spend time on the land in rich and fulfilling social and natural environs. In complementary fashion, barren-ground caribou seem to represent a harvestable species accessible to younger hunters who may lack expensive equipment like riverboats but who can arrange to get a snowmobile and some fuel. Similar to Ingold's (2000) account of the process of "enskillment," it seems that community hunts afford a means of stimulating such appropriate attention, whereas caribou hunts provide a mechanism to practise and develop the necessary skills for more technical hunting of moose, woodland caribou, and other species.

In relation to the decreasing availability of barren-ground caribou to communities such as Fort Good Hope, the results of this study suggest that young hunters may be particularly affected, as they are less able to pursue other game that requires additional expensive equipment and that may be more challenging to hunt in terms of the requisite knowledge and skill. This conclusion has implications for regional wildlife management organizations that may be contemplating various quota systems to manage caribou harvests (CBC North, 2009). In addition, the latest "caribou crisis" has led to the cancellation of some community hunts in the Northwest Territories (ibid.; GNWT, 2010), which could reduce opportunities for youth to learn from elders on the land. These hunts are part of the critical process of ensuring cultural continuity within northern Indigenous communities, and it is evident that such continuity is sought by residents of Fort Good Hope – young and old alike.

References

Asch, M. (1979). The economics of Dene self-determination. In D.H. Turner & G.A. Smith (Eds.), *Challenging Anthropology: A Critical Introduction to Social and Cultural Anthropology* (pp. 339–351). Toronto: McGraw-Hill Ryerson.

Auld, J. & Kershaw, R. (Eds.). (2005). *The Sahtu Atlas: Maps and Stories from the Sahtu Settlement Area in Canada's Northwest Territories*. Norman Wells: Government of the Northwest Territories.

Barnaby, G. (1976). *The Dene Political System: Summary of Evidence of George Barnaby before the Mackenzie Valley Pipeline Inquiry*. Yellowknife: Mackenzie Valley Pipeline Inquiry.

Barnhardt, R. (2005). Culture, community and place in Alaska Native education. *Democracy & Education, 16*(2), 59–64.

Barnhardt, R., & Kawagley, A. (2005). Indigenous knowledge systems and Alaska Native ways of knowing. *Anthropology & Education Quarterly, 36*(1), 8–24.

Bonny, E. (2008). *Inuit qaujimajatuqangit and knowledge transmission in a modern Inuit community: Perceptions and experiences of Mittimatalingmiut women* (MSc thesis). University of Manitoba, Winnipeg.

Brook, R. (2010). Collaborating with Indigenous communities in the Canadian Arctic to understand caribou form and function. Society for Advancement of Chicanos and Native Americans in Science (SACNAS), December 20. http://sacnas.org/about/stories/sacnas-news/fall-2009/collaborating-indigenous-communities.

CBC North (2009). N.W.T. First Nation cancels caribou hunt. *CBC News,* September 16. http://www.cbc.ca/news/canada/north/n-w-t-first-nation-cancels-caribou-hunt-1.819533.

Collings, P., & Condon, R. (1996). Blood on the ice: Status, self-esteem and ritual injury among Inuit hockey players. *Human Organization, 55*(3), 253–262.

Collings, P., Wenzel, G., & Condon, R. (1998). Modern food sharing networks and community integration in the central Canadian Arctic. *Arctic, 51*(4), 301–314.

Condon, R., Collings, P., & Wenzel, G. (1995). The best part of life: Subsistence hunting, ethnicity, and economic adaptation among young adult Inuit males. *Arctic, 48*(1), 31–46.

Davidson-Hunt, I., & Berkes, F. (2003). Learning as you journey: Anishinaabe perception of social-ecological environments and adaptive learning. *Conservation Ecology, 8*(1), 5.

Fajber, E. (1996). *The power of medicine: "Healing" and "tradition" among Dene women in Fort Good Hope, Northwest Territories* (MA thesis). McGill University, Montreal.

Ford, J., Pearce, T., Smit, B., Wandel, J., Allurut, M., & Shappa, K. (2007). Reducing vulnerability to climate change in the Arctic: The case of Nunavut, Canada. *Arctic, 60*(2), 150–166.

Gombay, N. (2009). Sharing or commoditising? A discussion of some of the socio-economic implications of Nunavik's Hunter Support Program. *Polar Record, 45*(233), 119–132.

Government of the Northwest Territories (GNWT) (2000). *Community Wellness in Action 1998–1999: Summary Report of Community Wellness Initiatives*. Yellowknife: Department of Health and Social Services, Government of the Northwest Territories. http://pubs.aina.ucalgary.ca/health/62291.pdf.

Government of the Northwest Territories (GNWT) (2010). Yellowknives Dene First Nation and GNWT sign agreement to support recovery of Bathurst caribou herd (Press Release). October 7. http://www.enr.gov.nt.ca/sites/default/files/yellowknives_dene_and_gnwt_sign_agreement.pdf.

Hall, E. (Ed.). (1989). *People and Caribou in the Northwest Territories*. Yellowknife: Department of Renewable Resources, Government of the Northwest Territories.

Helm, J. (1965). Patterns of allocation among the Arctic Drainage Dene. In J. Helm, P. Bohannan, & M. Sahlins (Eds.), *Essays in Economic Anthropology: Proceedings of the 1965 Annual Spring Meeting of the American Ethnological Society* (pp. 33–45). Seattle: American Ethnological Society.

Hickey, C., Nelson, M., & Natcher, D. (2005). Social and economic barriers to harvesting in a northern Alberta Aboriginal community. *Anthropologica, 47*(2), 289–301.

Hurlbert, J. (1962). *Age as a Factor in the Social Organization of the Hare Indian of Fort Good Hope, N.W.T.* Ottawa: Northern Coordination and Research Centre.

Ingold, T. (2000). *The Perception of the Environment: Essays in Livelihood, Dwelling and Skill*. London, New York: Routledge.

Johnson, M. (1992). Dene traditional knowledge. *Northern Perspectives, 20*(1), 3–5.

Johnson, M., & Ruttan, R. (1993). *Traditional Dene Environmental Knowledge: A Pilot Project Conducted in Ft. Good Hope and Colville Lake, N.W.T., 1989–1993*. Hay River: Dene Cultural Institute.

Kawagley, O., & Barnhardt, R. (1999). Education indigenous to place: Western science meets native reality. In G.A. Smith & D.R. Williams (Eds.), *Ecological Education in Action: On Weaving Education, Culture and the Environment* (pp. 117–140). New York: State University of New York Press.

Kishigami, N. (2000). Contemporary Inuit food sharing and hunter support program of Nunavik, Canada. In G. Wenzel, G. Hovelsrud-Broda, & N. Kishigami (Eds.), *The Social Economy of Sharing: Resource Allocation and Modern Hunter-Gatherers* (pp. 171–92). Kyoto: Senri Ethnological Studies.

Kishigami, N. (2004). Contemporary Inuit food sharing: A case study from Akulivik, PQ, Canada. Paper presented at the conference Effective Local Institutions for Collective Action in Arctic Communities, Fifth International Congress of Arctic Social Sciences, University of Alaska, Fairbanks.

Kulchyski, P. (2005). *Like the Sound of a Drum.* Winnipeg: University of Manitoba Press.

Kumpulainen, K., & Renshaw, P. (2007). Cultures of learning. *International Journal of Education and Research, 46*(3–4), 109–115.

Nadasdy, P. (2005). The anti-politics of TEK: The institutionalization of co-management discourse and practice. *Anthropologica, 47*(2), 215–232.

Nelson, M. (2005). Paradigm shifts in Aboriginal cultures? Understanding TEK in historical and cultural context. *Canadian Journal of Native Studies, 25*(1), 289–310.

Nuttall, M., Berkes, F., Forbes, B., Kofinas, G., Vlassova, T., & Wenzel, G. (2005). Hunting, herding, fishing, and gathering: Indigenous peoples and renewable resource use in the Arctic. In C. Symon, L. Arris, & B. Heal (Eds.), *Arctic Climate Impact Assessment* (pp. 649–690). New York: Cambridge University Press.

Ohmagari, K., & Berkes, F. (1997). Transmission of Indigenous knowledge and bush skills among the western James Bay Cree women of Subarctic Canada. *Human Ecology, 25*(2), 197–222.

Parlee, B., Berkes, F., & Teetl'it Gwich'in Renewable Resources Council. (2006). Indigenous knowledge of ecological variability and commons management: A case study on berry harvesting from northern Canada. *Human Ecology, 34*(4), 515–528.

Pearce, T., Wright, H., Notaina, R., Kudlak, A., Smit, B., Ford, J., & Furgal, C. (2011). Transmission of environmental knowledge and land skills among Inuit men in Ulukhaktok, Northwest Territories, Canada. *Human Ecology, 39*(3), 271–288.

Peloquin, C., & Berkes, F. (2009). Local knowledge, subsistence harvests, and social-ecological complexity in James Bay. *Human Ecology, 37*(5), 533–545.

Salokangas, R. (2009). *The meaning of education for Inuvialuit in Tuktoyaktuk, NWT, Canada* (MSc thesis). University of Alberta, Edmonton.

Savishinsky, J. (1974). *The Trail of the Hare: Life and Stress in an Arctic Community.* New York: Gordon and Breach Science Publishers.

Savishinsky, J., & Hara, H. (1978). Hare. In J. Helm (Ed.), *Handbook of North American Indians,* Volume 6, *Subarctic* (pp. 314–25). Washington, DC: Smithsonian Institution Press.

Usher, P.J. (2000). *Standard Edible Weights of Harvested Species in the Inuvialuit Settlement Region.* Inuvik: Wildlife Management Advisory Council.

Wenzel, G. (1995). Ningiqtuq: Resource sharing and generalized reciprocity in Clyde River, Nunavut. *Arctic Anthropology, 32*(2), 43–60.

Wray, K. (2011). *Ways we respect caribou: Hunting in Teetl'it Zheh (Fort McPherson, NWT)* (MSc thesis). University of Alberta, Edmonton.

PART 3
Food Security

8
Time, Effort, Practice, and Patience

Anne Marie Jackson

First Nation peoples have always been culturally connected to the land. We were once a nomadic people. Certainly, times have changed and, with them, our culture and traditions. The concern about our culture and traditions is that they are being lost – unpractised. Our elders continue to stress this message. But they also emphasize the importance of education. Youth, in turn, are faced with learning to balance their culture and traditions with today's modern society. From our history stems our cultural identity; our history is the crucial ingredient that sets us apart from other ethnic groups, and we must know our history to understand and appreciate our unique traits.

The changes throughout our history have taken a toll on First Nation cultures and traditions. Our elders lived a nomadic life before their sudden transition to communities and houses. With this transition came a rapid push to assimilate our First Nation peoples, stripping us of our cultural and traditional ways. But we have not completely lost our heritage. Our elders still retain their knowledge and oral teachings. They hold their knowledge dear and wish to promote it and see it practised.

The fight to keep our culture and traditions alive seems to be unending. Sometimes there is too much focus on being part of mainstream society and not enough on our culture and traditions or vice versa. The struggle is to find balance in both worlds. Our eldest elders of the small communities in the North, who lived entirely on the land before they were relocated to communities and placed in houses, realize and stress the importance of sustaining our heritage; after all, who are we without our culture and traditions?

Concerned about youth losing their cultural roots because aspects of the culture are less practised, elders and community leaders are taking the initiative to protect those ancient teachings, techniques, and practices. In our community of Fort Good Hope, community organizations have formed On the Land or Back to the Land programs to provide local employment in the face of significant unemployment and poverty but also, most importantly, to instil culture and traditions into the youth. Our local school also tries to maintain our heritage in indoor and outdoor classroom settings. Community hunts are another opportunity for elders to come together and teach youth. There is certainly always room for more innovative techniques to propel culture and traditional practices.

Fortunately, like my siblings, I grew up very close to our history and culture from when I was an infant to my teens with my mother and father, as well as my grandpa and grandma; to them, our history was part of life. I was interested and wanted to learn. This is how my family lived. I admired the skills they had for surviving on the land and the stories they shared of their lives and their parents' lives. They taught me to understand the environment from its vegetation, as well as its inhabitants and their migration routes. They showed me the skills of living from the land so that I could harvest and provide food and could understand and adapt to seasonal changes. Both of my parents took the time and effort to patiently pass on our culture and traditions to my siblings and me, as well as to support us in developing the skills required to integrate into society. Having lived that lifestyle, I am motivated to continue it throughout my life and eventually to share it with my children and to promote it to others.

I caught my first rabbit when I was fourteen. My family and I made camp along the shore of the Mackenzie River. I thought, "Okay, I want to go do this by myself. I want to show I can catch a rabbit, too." I took my dog with me, and I was very scared. I walked to the end of the shoreline that made way for a creek or river. I thought it was a good place because there were tall willows. I set one snare. I woke up the next morning anxious to check my snare, so I went back to check. And I had caught a rabbit. My father was very proud of me, and I was proud of myself because I had done something without relying on my father to be there. It made me feel like I could do this on my own now. So the next time he brought me out to the bush – I went to Manuel Lake with him – I told him, "I'm going to set snares," and he said, "Go ahead."

I feel good about myself because I can provide.

Our eldest elders have made the rapid transition from a nomadic life to community life, and they understand the importance of "importing" today's society. After all, how can we survive and advance in society if we continue on with just our traditional way of life? But it is important for First Nation people not to completely forget our true identity and to take time and effort to practise our traditions.

My father lives the life of a hunter, trapper, and gatherer, but he has also started a tourist business in his homeland. I would like to think he chose that career because he knows the land and wants to share with tourists the importance of our identity. He stresses the importance of both worlds but puts our culture and traditions first and foremost. He has found his balance. My mother, a political force through her devotion to leadership, fights to sustain and protect the land and all its contents. She has never ceased to emphasize the importance of education, as well as to use it to the advantage of First Nations so that our cultural and traditional practices are part of the current governing system. Through our lifestyle and example, we are living proof that these practices can be sustained. But we also live with the struggles that come with this effort. First Nation peoples value respect, balance, and harmony in our culture and traditions, and these values conflict with today's social and economic structure. Most importantly, if our culture and traditions are important to our First Nation identity, then it starts with us to practise them now.

9

The Wage Economy and Caribou Harvesting

Zoe Todd and Brenda Parlee

In achieving food security, northern communities are faced with many kinds of stresses and influences (Duhaime & Bernard, 2002; Duhaime & Caron, 2009; Southcott, 2015). Much food security research has focused on mapping or describing dietary patterns, with less consideration given to the influence of ecological variabilities, including those associated with the most recent decline in barren-ground caribou. However, linear connections between caribou population decline, on the one hand, and food security outcomes, on the other, cannot easily be drawn. There are many social, economic, and political factors that influence how communities cope with and adapt to the changing availability of traditional/country foods. In this analysis, we consider how wage employment, coupled with harvest regulation by regional co-management boards, mediates and compounds the impact of a declining Cape Bathurst and Bluenose West caribou population in northern Canada. In using this broader socio-economic and political lens to interpret the observations and experiences of Paulatuuqmiut (Paulatuuq people), we aim to address the question of how caribou population decline affects food security in the western Arctic.

BACKGROUND

"Food security [is] a situation that exists when all people, at all times, have physical, social and economic access to sufficient, safe and nutritious food that meets their dietary needs and food preferences for an active and healthy life" (Food and Agriculture Organization, 2015, p. 53). The importance of traditional foods, such as caribou, fish, beluga, muskox, and geese, to food security in the Arctic is well established (Lambden, Receveur, & Kuhnlein, 2007; Willows, 2005; Usher, 2002). The availability of traditional foods in remote communities is particularly important to food security given that market foods are higher in cost, limited in quantity, and poorer in quality and can be culturally inappropriate. But the barriers to traditional food harvesting in many parts of northern Canada appear to be growing (Larsen & Huskey, 2015).

Contemporary theorizing on food security goes beyond calculation of costs and nutritional attributes. There is greater attention paid to the subjectivity of the process of achieving security and to the diverse meanings that might be associated with its outcome (Mintz & Du Bois, 2002; Pottier, 1999). For many Indigenous peoples, food insecurity is a lived experience that reflects the social, economic, and cultural realities of daily life, including the availability of food resources. For Indigenous communities, who seek to maintain strong traditional food diets and economies, food security is a concept interwoven with the complex histories of place, legal rights to access resources, as well as the overall health and availability of those resources (Mares & Peña, 2011; Power, 2008).

Both store-bought and traditional foods play an important role in the northern food security equation; for northern and remote communities with strong links to their local environment and limited market food options, traditional foods play a critical role in physical health and well-being (Power, 2008). This is particularly true in the community of Paulatuk, Northwest Territories.

Notions of food security cannot be rigidly calculated or applied in dynamic Arctic environments. Uncertainty in the availability of Arctic marine and terrestrial food is characteristic of Arctic environments (Chapin et al., 2004). Historically, many northern communities were well adapted to the comings and goings of fish, wildlife, and other resources (Nuttall et al., 2005; Peloquin & Berkes, 2009). However, many factors are increasingly complicating natural patterns of variability and limiting the capacity of individuals and communities to access traditional food sources. Among these factors are the increasing presence of contaminants as well as the impacts of climate change (Donaldson et al., 2010; Ford, Pearce, Duerden, Furgal, & Smit, 2010; Paci, Dickson, Nickels, Chan, & Furgal, 2004). Large-scale resource development projects are also having a profound effect on food harvesting across the circumpolar North, including northern Canada (Duhaime & Caron, 2006).

Resource development is also altering other kinds of social-ecological relations (Duhaime & Caron, 2006). There is a large body of work on the mixed economy that suggests increasing opportunities for wage employment are adversely affecting the amount of time spent harvesting in the short term and the persistence of knowledge and skills for harvesting in the medium to long term (Kruse et al., 2008; Langdon, 1986; Usher, Duhaime, & Searles, 2003). A comparable body of work suggests there is a beneficial impact on harvesting. Research in the Northern Slope of Alaska demonstrates how employment in the oil and gas sector influences harvesting by increasing opportunities to harvest among those who have well-paying jobs (Kruse, 1991). In the North Slope, an area with a history of oil and gas activity, Kruse (ibid., p. 320) found that "forty-five percent of households with incomes of $60,000 or more in 1988 reported that over half their food came from subsistence, a higher proportion than that reported by any other income group." This parallels research in Canada. Condon, Collings, and Wenzel (1995, p. 37) found "the most active hunters in our sample population are also those who have regular (and in many cases, high-paying) jobs which provide enough income for the purchase of equipment and supplies." Advances in transportation and communication technology (e.g., snowmobiles, four-wheel-drive all-terrain vehicles, and air travel) are theorized as having further beneficial effects, as individuals have the means to be more efficient in their land-use activities (Condon et al., 1995). These findings may be strictly local, although similar research findings were obtained during food security research in Paulatuk (Todd, 2010). Larger-scale studies such as the Survey of Living Conditions in the Arctic (SLiCA) found that there was neither a positive nor a negative relationship between wage work and subsistence activities (Kruse et al., 2008, p. 116). The relationship between employment and harvesting is thus complex and nuanced, with local factors playing a role in shaping the relationship between both activities (Berman & Kofinas, 2004; Chabot, 2003; Condon et al., 1995; Kerkvliet & Nebesky, 1997; Kruse, 1991; Mackenzie Gas Project, 2004; Nadasdy, 2003; Stabler, 1990).

Further complicating these relationships is the influence of government or regulatory agencies involved in managing harvest activities. Historically, government had a significant impact on harvesting; up until the 1980s, there was a tendency to criminalize or deter Aboriginal participation in subsistence practices, including caribou harvesting (Sandlos, 2007). The settlement of land claims, including the Inuvialuit Final Agreement, and consequent creation of co-management boards to protect and govern harvesting, has significantly changed the landscape of natural resource management in northern Canada; however, there are still numerous concerns and conflicts between policies and regulatory instruments aimed at natural resource conservation and those aimed at ensuring the continuation of cultural harvesting practices of Aboriginal peoples.

In this study, we are particularly concerned with the intersection between three problems or realities: the availability of barren-ground caribou, changes in wage employment opportunities, and the regulation of harvesting by co-management boards.

SETTING, APPROACH, AND METHODS

The chapter is based on research carried out in Paulatuk in 2008, during which semi-directed interviews were conducted with numerous harvesters and nonharvesters to assess how wage employment impacts traditional harvesting, dietary patterns, and social networks in the Inuvialuit Settlement Region. Paulatuk is a hamlet situated in that region and is located roughly 400 kilometres northeast of Inuvik. Nestled at the base of Cape Parry, in Darnley Bay, it is home to 329 community members (NWT Bureau of Statistics, 2012, p. 1). Historically, the region around Paulatuk was important to Inuvialuit people's Thule ancestors, and more contemporarily the Igluyuaryungmiut used the lands around Paulatuk very extensively prior to the arrival of non-Indigenous whalers and traders (Alunik, Kolausok, & Morrison, 2003, p. 17). This historic land use and occupation are evident at archaeological sites within Tuktut Nogait National Park, which is situated to the east of the community.

Throughout the 1920s and 1930s, a number of local families continued to hunt, fish, and trap along the coast of the Beaufort Sea between the Horton River to the west and Pearce Point to the east, taking advantage of fur-trading opportunities at Hudson's Bay Company posts at Baillie Island and Letty Harbour, while accessing goods from Catholic missions built at the base of Darnley Bay and Letty Harbour (McDonnell, 1983, pp. 34–55; Usher, 1971, pp. 174–80). In the 1950s a Distant Early Warning (DEW) Line site was built on the tip of the Parry Peninsula, and some families relocated their homes to Cape Parry in order to take advantage of associated wage opportunities. However, families struggled to access adequate caribou, fish, and other foods at Cape Parry (Abrahamson, 1963, pp. 25–28), and in 1967 local leaders established a new community site at Paulatuk's current location, which enabled better access to abundant animal, fish, and plant resources. Today, the community is governed under the Inuvialuit

Final Agreement, and Paulatuuqmiut participate in harvesting activities as well as wage opportunities. These wage opportunities include work with Inuvialuit self-government, infrastructure and governance jobs with the hamlet, as well as employment in education, health, and wildlife monitoring and management. There is strong interest in potential nickel and diamond mining close to the community (Darnley Bay Resources Ltd., 2011).

The interview guide employed a mixture of quantitative and open-ended qualitative questions. Twenty participants ranging in age from twenty to eighty-three were interviewed with the intention of identifying an equal number of men and women participating in the wage economy full time, part time, and seasonally, as well as those not currently employed. Eleven men and nine women were interviewed following a "snowball" or "chain" sampling method (Creswell, 2007, p. 127). Education, gender, age, and length of time residing in the community were accounted for in the selection of participants. Of the nine women interviewed, two were retired, one worked seasonally, two worked part time, and four worked full time. Of the eleven men interviewed, three were unemployed, one worked a rotational job outside of the community, three worked seasonally, three worked part time, and one worked full time. Given the small sample size, it was not possible to control for all of these factors or to create a perfectly representative sample. In the summer of 2009, a follow-up workshop was held on food security with eleven separately recruited participants. This chapter focuses on the findings from a literature review, the twenty interviews, and the workshop.

RESULTS

Changes in Caribou Population

The community of Paulatuk is dependent upon the harvest of diverse terrestrial and marine species (Community of Paulatuk et al., 2008). The main caribou herds being harvested by Paulatuuqmiut are the Cape Bathurst, Bluenose East, and Bluenose West, which are referred to collectively as tuktu by the Paulatuuqmiut (see Figure 9.1). Although these three herds are currently defined and managed separately, the Government of the Northwest Territories managed them as one for thirty years as the "Bluenose Caribou Herd."

Aerial surveys have produced population change statistics for both the Bluenose West and Cape Bathurst herds. The lack of differentiation of the herds prior to the 1990s affected the way that statistics on population were gathered, thus limiting a longitudinal perspective on population change (see Figure 9.2).

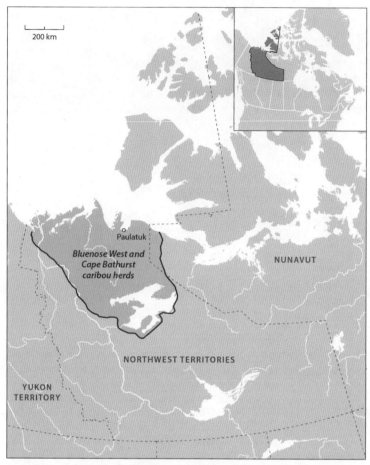

FIGURE 9.1 Location of the Bluenose West and Cape Bathurst caribou herds

Based on current aerial surveys by the Government of the Northwest Territories, it is estimated that the Cape Bathurst herd declined from 20,000 animals in 1992 to about 2,000 animals in 2005 and 2006 (Nagy, 2009). The 2009 population estimate showed the herd to have been stable since 2006. The Bluenose West herd reportedly declined from an estimated high of over 110,000 animals in 1992 to about 18,000 by 2006, with numbers seemingly stabilizing at this level in 2009 (Environment and Natural Resources, 2010; Nagy, 2009). Although these population surveys have been critiqued (see Chapter 3), these changes in numbers are proposed as affecting the numbers of caribou harvested and consequently the availability of traditional/country foods. However, ecological conditions only partly explain the issue.

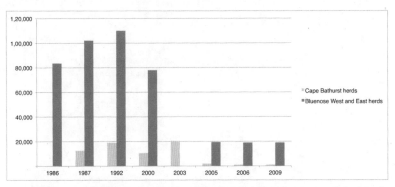

FIGURE 9.2   Population estimates for the Bluenose West, Bluenose East, and Cape Bathurst herds, 1986–2009
Source: Adapted from Davison (2016).

Changing Institutions: Harvest Regulation

Reports of decreasing herd numbers led co-management boards and the Northwest Territories government to enter into discussions of harvest management. In 2007 a total allowable harvest of 4 percent of the estimated 2006 Bluenose West herd was implemented (Environment and Natural Resources, 2011). This figure translates into substantially lower local harvests than the averages calculated during the period of the Inuvialuit harvest study, which were between four and nine caribou per household in Paulatuk (Inuvialuit Joint Secretariat, 2003). A similar quota applies to hunting in regions S/BC/01 and S/BC/03, which are also areas where Paulatuuqmiut have traditionally harvested caribou.

Changes in Employment

The economy in Paulatuk is characterized by a mix of wage employment and traditional economic activities. The majority of individuals employed in wage employment in Paulatuk work full time (72.4 percent) (see Table 9.1).

Participation in the traditional economy through harvesting remains an important aspect of life in Paulatuk. In 2003, 49.5 percent of individuals in Paulatuk hunted and fished, and 51.9 percent of households consumed country foods, defined as "most or all (75% or more) of the meat or fish consumed" in the household (NWT Bureau of Statistics, 2007, p. 2). These values are much higher than those in the Northwest Territories as a whole (see Table 9.2).

TABLE 9.1 A comparison of employment characteristics in Paulatuk and the whole of the Northwest Territories

	Paulatuk (%)	NWT (%)
Labour force participation rate (2006)	58.1	76.5
Unemployment rate (2006)	28	10.4
Employment rate (2006)		
Females	36.8	66.7
Males	43.5	70.1
Aboriginal	37.8	52.2
Non-Aboriginal	60	82.8
Potential labour supply (2006)		
Unemployed (people)	31	2,454
Do rotational work	87.1	70.3
Males	71	64.4
Employment profile (2006)		
Full time	72.4	85.9
Part time	19.4	11.6
Government, health, social services, education	48	37.3
Goods producing	8	17.2
Other industries	48	43.9
Annual work pattern		
Worked in 2005	62.8	81.2
Worked more than 26 weeks	51.9	75.5

Source: NWT Bureau of Statistics (2010, pp. 3–4) and for figures on full-time and part-time employment, NWT Bureau of Statistics (2007, p. 3).

TABLE 9.2 A comparison of harvesting activity and consumption of country foods in Paulatuk and the whole of the Northwest Territories, 2003

	Residents hunting and fishing (%)	Households consuming country foods
Paulatuk	49.5	51.9
Northwest Territories	36.7	17.5

Source: NWT Bureau of Statistics (2007, p. 2).

Wage employment opportunities are anticipated to be on the increase. Paulatuk is the site of potential regional impacts from oil and gas activity with the proposed Mackenzie Gas Project (2004), as well as the site of direct local impacts from mineral exploration and proposed nickel and diamond mining projects (Diadem Resources Ltd., 2011; Keeping, 1999). The authors of a socio-economic impact assessment of the Mackenzie Gas Project (2004, p. 6-19) anticipate that "the well-being of individuals and groups can benefit because of employment opportunities and project spending that will provide increased income to spend on improving quality of life in the community or harvesting on the land." The authors further suggest that benefits will accrue from the fact that "harvesting and seasonal employment are now symbiotic, because low incomes from trapping necessitate wage employment to pay for the expensive equipment now needed for efficient harvesting. The project will provide wage employment that will support harvesting-equipment requirements" (ibid., p. 6-28). However, the authors also anticipate that employment in the project may have negative impacts on traditional harvesting: "Project employment could jeopardize harvester lore and disciplines by bringing Aboriginal and non-Aboriginal workers together on the job, and by preempting harvesting activities, because of time needed for long-rotation employment cycles. Some Aboriginal people might experience the paid work more rewarding than harvesting, promoting interest in a southern lifestyle" (ibid., p. 6-29).

Although the latter statement overly simplifies the relationship between harvesting and employment, what is clear from research in Paulatuk is that the relationship between harvesting and contemporary employment associated with proposed oil and gas or mining activity in the community may be nuanced, with both benefits and drawbacks. The amount, structure, and value of time as a determinant of participation in harvesting activities is acknowledged in the socio-economic impact assessment of the Mackenzie Gas Project (2005, pp. 7-3, 7-5):

> The project will affect traditional harvesting through effects on the relevant time and resources available to Aboriginal people for harvesting, and on their motivation to do the harvesting work. Large project demands for workers, and a range of employment opportunities, will be found throughout the study area, including Paulatuk ...
>
> Measures that will be undertaken by the project proponents include:
>
> - providing flexible work schedules to accommodate traditional harvesting and other Aboriginal cultural, family and community needs, where

practical, recognizing that work flexibility will be limited in the peak winter construction seasons.

Beyond the impact of the Mackenzie Gas Project, Paulatuk currently faces potential mineral development that would bring employment to the community (Darnley Bay Resources Ltd., 2011; Diadem Resources Ltd., 2011; Keeping, 1999). In the face of regional projects like the Mackenzie Gas Project and local projects such as the Darnley Bay Nickel Mine, the influence of the wage economy on harvesting and food security takes on special significance, as the community and territorial and federal governments have the opportunity to set guidelines that encourage project proponents to structure jobs in such a way as to acknowledge the relationships at play in the mixed economy of Paulatuk.

Changes in Harvesting

The practice of traditional food harvesting is fundamental to residents of Paulatuk. Interview participants' responses focused mostly on geese hunting, caribou hunting, and fishing in the spring; on fishing, caribou hunting, and beluga hunting in the summer; and on fishing and caribou hunting in the fall. One participant, Neve, described how the year is spent:

And our time out on the land depends on our animals, mainly. Geese and other birds come in the spring, so we're out then. And for the men, the more brave ones, they go in May, spring hunt. June, mid- to late, mid-June, the men are out on the ice, doing their char fishing. July, caribou hunting. July, August, char and caribou hunting, September as well. Well, between July and September, caribou, char, even go after seals, whales. August and September, we try to get out, too. And, of course, if we have no holiday time, we're out in the weekends, the summer months, berry picking, caribou hunting, fishing.

There is, overall, limited activity in the winter months for most participants, although some indicated that they currently harvest in the winter, and some indicated that they do go out on the land in the winter if they absolutely have to. In the past, when the trapping economy was still strong, winter harvesting was a more common activity. As Bob explained,

That was in the '70s. That was the good years. And wolves were some ranging between $250 to $300. People were always enjoying it out on the land.

It didn't matter how cold it was. They were hardy people. I remember one
winter it was so cold when you travel with your skidoo, you can see the
smoke of it for miles, it just hanging there, but we didn't feel it. We were
used to it. Nowadays, you hardly see people travel up inland to do any
trapping or anything. You know. There's a few hardy people, just a handful,
that go out during the month of January, February. Oh, we do polar bear
hunts, also, during the winter.

Although harvesting activity has changed over the years, it remains
an important aspect of life in Paulatuk, and the traditional economy is
vibrant and viable. In fact, a majority of the sixteen participants indicated
that they would like to be able to spend more time on the land, but
many indicated that employment and time are two of the biggest factors
that impede their ability to spend more time on the land. As Melanie
explained,

Yeah. It would be, we'd be able to camp a few days on the land, and we'd
be able to do more things, like learning how to pluck geese, or cut fish, or
stuff like that, cut up caribou. And I'm glad I know how to do all that just
from learning from my mom. And, well, it was more enjoyable to know
that you don't have to rush back for work and stuff like that, but now, it's
just I can't really get time off for going out on the land unless I take time
off without pay, and that's something I can't really do with my family. But
it's good 'cause I have my sisters and brothers that do the hunting.

This account reveals an important relationship between harvesting, time
spent with family, and learning. Harvesting activities are not just economic pursuits but also social and cultural activities embedded within
a complex socio-economic world. The benefits of harvesting and participation in the traditional economy go beyond the utility of procuring
food since these activities play an important social and cultural role and
are an important factor in personal well-being. For Bob, harvesting is
a "family affair" that brings people together on the land that stretches
around the "four corners of Paulatuk – north, south, east, and west."
As he explained,

The summer? It's another family affair. You, they go out. It's about ten
miles from here, Argo Bay, Green's Island. They do their fishing there.
Whenever caribou go passed by, you get it. Basically, fishing and enjoying the July month. Also, there's beluga hunts during the month of July.

[Pause] August, you have your char fishing. That's another family affair. You go out, camp at the river.

Not everybody is able to go out on the land with their families. However, those who cannot get out on the land still appreciate it when other family members can get out and travel. For example, Melanie is happy that her brothers, sisters, and son are able to enjoy time on the land and provide country foods for the family:

> I mean, like, this land around Paulatuk is so beautiful, but I don't get to enjoy it as much as I used to. Like, I'll be lucky if I go out for a day trip, so – and I'm so glad that my son is going out geese hunting and caribou hunting and stuff. He's already getting ready to go out this weekend for geese hunting. They're sighting geese coming, so they're excited.

Harvesting also connects individuals to family members who have passed away; participants spoke about their connection to the land through memories of past harvesting trips and how harvesting also conjures memories of deceased family members or brings families together after someone has passed away. As Neve explained,

> One spring hunt, my parents were both gone, both died, and that spring lot of, most of my family members were out, my brothers and my sisters. The whole family, and husbands and wives, were out fishing. It was a beautiful day, just enjoying ourselves, it was. And one of my brothers said that this is the life, that's what he said. So that made it really special.

Neve noted that participation in the traditional economy also offers individuals an opportunity to get out of town and take a holiday:

> Because we like going out on the land so much, the free time I have, instead of going down south with the family – for one thing, it costs too much to go down there, and then not all of you go out, but with on the land travel, the whole family can go out. Everything is close by. It doesn't cost that much to go out.

For Neve, harvesting is thus an opportunity to reconnect with family. Whereas outsiders may see harvesting only through an economic lens, for community members, it takes on many interdependent and mutual roles.

As in many other communities in the western Arctic, caribou are seen as critical to the well-being of the Inuvialuit people. It is well established that because of its major nutritional value, caribou meat is important to their diet. But as noted by Janet, caribou meat also has therapeutic and medicinal value:

> You know, you hear about this old wives' tale on chicken soup; up here it's the broth from caribou – caribou head or caribou broth that is almost a medicine for anyone who is sick or wanting to get better. It's just like a good, healthy dose of Tylenol. Even better, it's the nutritional dose of vitamins that your body needs to get strong again. You'll never find it in chicken soup. You'll find more chemicals in chicken soup than there is in a box of – in a cardboard box.

There are many factors that influence caribou harvesting in the community of Paulatuk. The decline of caribou is a key concern. However, despite reports from aerial surveys that caribou were in decline, local harvesters in Paulatuk did not universally perceive a decrease in the availability of caribou. Essentially, a paradox of science has arisen in that people in the community are seeing caribou but being told they are not there.

In 2007 a caribou quota that was proposed by the regional Inuvialuit Game Council was adopted by the Paulatuk Hunters and Trappers Committee. This measure was in response to wildlife surveys suggesting that the Bluenose West caribou herd, which the community relies on for its harvests, was in decline (Environment and Natural Resources, 2010). Interview responses, as well as discussion during the food security workshop, highlighted how restricted access to the herd is impacting access to nutritious and culturally appropriate food in the community. As Janet explained,

> I don't think it's going to take very long to harvest our two meagre caribou for – and for a community like this with a 300 percent inflation rate, even full-time employees have to go out on the land to harvest to sustain their families for the whole year. So then it's the summer, and this last year it was curing and putting away the beluga whales, and then in August the Arctic char fishing up at the mouth of the Hornaday River, and September and October, camping right inside the river.

Michael, a full-time employee, pointed out that the quota and other wildlife regulations are impacting access to country foods in the community:

> Now, with the quota system on the caribou, you can't do what you could do at one time. We're allowed two caribou for the year, whether you get it in the spring or fall. So, you know, the effect to our ability to harvest is not only job-related: it's also related to legislation, our new legislation, and the quota system. And if they ever quota our geese, man, we're cooked. We've got a quota on our char, we've now got a quota on our caribou.

Respondents indicated that they are trying to replace caribou meat with store-bought meats and other animals. However, as Dorothy illustrated, this approach can be challenging: "Because they don't get caribou this year, eh? We have no meat. We buy lots of meat from the store now, but expensive. But we have to eat. When there's lots of caribou, you don't have to buy that much." Neve pointed out that the quota causes her to buy more store-bought meat: "This past year. Well, since, I'm not sure how many months, but since the limit on caribou came up, then we're buying more meat products." There are opportunities to substitute caribou with other foods from the land; however, as Donald pointed out, these options are dwindling: "Yes, we harvest a lot of – in the springtime we try and get some caribou, too. But now with our caribou, we only get two tags per year, so now we have to wait until next year to get our two tags. That's kind of hard, so we're going to start doing a lot of fishing." Bob explained, "Right now, our caribou is changed. We can't harvest as much as we want, but the geese never change. We can get as many as you like. As long as you don't run out of bullets, just keep going. After you've harvested, you always share it out."

John emphasized that in addition to losing the product of the harvest, a quota also means that the experience of harvesting is lost or changed:

> Fall. I like fishing and caribou. Mostly, you know, freeze them and gather them up. Can't do that much now. There's so many limits now. Can't dry [fish or caribou meat] – can't do that now [since there is not enough meat]. So I got to do other things now to survive, so fish and caribou and the ones that got no restriction on them. Polar bear, seals, and fish, and that's what I got to do now. Can't get much caribou.

John thus demonstrated how the quota is altering the types of harvesting activities he pursues. This quota is also impacting the sharing of country foods in the community since there is less food for harvesters to share. As John pointed out,

> Oh, about three years ago, I get a 175 [caribou]. That's just for my whole family, the whole bunch of us here. But they still don't last all winter. There's a whole a bunch of us, that's why. Barely made it to April, that time, even [though] I get 175 now in September ... Well, we're allowed two. That's it now. Can't do too much now.

This account indicates how quickly the quota was felt among sharing networks within the community. Christine illustrated that the quota is also affecting the shape of collective harvesting activity:

> Caribou hunt, everybody participates. They camp by days and weeks. And it's gonna be a different amount of caribou because, you know, we're limited now. All we have to go, and through tags. It's two tags per person or household. Usually, on the fall time, my crew of boys participates. We participate through cutting, and butchering, and storing.

The regulatory changes are thus directly impacting how people interact with one another both on the land, where caribou are harvested, and back in the community, where caribou are processed and shared. This consequence merits the future attention of regulators and researchers.

* * *

The impact of declining barren-ground caribou populations on food security is a critical concern throughout the western Arctic and elsewhere. The impacts are not even or homogenous within or across communities and regions. One of the key factors affecting the significance of the impact is time spent in harvesting, a practice increasingly affected by the ups and downs in wage employment. In Paulatuk, harvest regulations put in place in 2007 reduced the allowable harvest to two caribou per household. The impacts of wildlife regulations on food security and on health and well-being have not been examined broadly. Duhaime et al. (2008, p. 79) point out that international declarations such as the Rovaniemi Declaration and Agenda 21 recognize that "Aboriginal access to land and land resources is considered a key part of the food security strategy, a means of promoting cultural activities and traditional knowledge, and a sustainable development aim for the Arctic." However, although the right to harvest is protected by legal frameworks, Paulatuk's experiences demonstrate how wildlife regulations can still have an impact on access to animals that Inuvialuit rely on for food. The views expressed by participants in this research illustrate that wildlife regulations should be examined in terms of their social and health impacts on a community as well as how they affect wildlife. The interactions between harvesting, the wage economy, wildlife regulations, and food security should thus be examined together, or holistically. Caribou regulations not only affect the animals in a herd but also intimately affect the harvesters who rely on them. Regulatory changes have the possibility to impact livelihood strategies, and this consequence merits greater attention in both regulatory and research approaches. Wildlife regulations add another dimension to food security concerns in Paulatuk. Although this study shows that the wage economy influences harvesting activity, the cumulative impact of other factors, such as wildlife regulations, also shapes how Paulatuuqmiut are able to access food from the land.

As the community plans for increasing resource development activity and contemplates the implications for community well-being, including food security, this chapter suggests the importance of dealing with the ups and downs in traditional/country food harvesting associated with wage employment, harvest regulation, and changing caribou populations.

REFERENCES

Abrahamson, G. (1963). *Tuktoyaktuk–Cape Parry: An Area Economic Survey, 1962*. Ottawa: Queen's Printer.

Alunik, I., Kolausok, E., & Morrison, D. (2003). *Across Time and Tundra: The Inuvialuit of the Western Arctic*. Seattle: University of Washington Press.

Berman, M., & Kofinas, G. (2004). Hunting for models: Grounded and rational choice approaches to analyzing climate effects on subsistence hunting in an Arctic community. *Ecological Economics, 49*(1), 31–46.

Chabot, M. (2003). Economic changes, household strategies, and social relations of contemporary Nunavik Inuit. *Polar Record, 39*(208), 19–34.

Chapin, F.S., Peterson, G., Berkes, F., Callaghan, T.V., Angelstam, P., Apps, M., . . . Whiteman, G. (2004). Resilience and vulnerability of northern regions to social and environmental change. *Ambio, 33*(6), 344–349.

Community of Paulatuk NWT Wildlife Management Advisory Council, & Inuvialuit Joint Secretariat. (2008). *Paulatuk Community Conservation Plan*. Inuvik: Community of Paulatuk. http://www.screeningcommittee.ca/pdf/ccp/Paulatuk_CCP.pdf.

Condon, R., Collings, P., & Wenzel, G. (1995). The best part of life: Subsistence hunting, ethnicity, and economic adaptation among young adult Inuit males. *Arctic, 48*(1), 31–46.

Creswell, J. (2007). *Qualitative Inquiry and Research Design: Choosing among Five Approaches*. Thousand Oaks, CA: Sage.

Darnley Bay Resources Ltd. (2011). *Darnley Bay Resources Project*. Accessed 2013. http:// www.darnleybay.com/index.aspx.

Davison, T. (2016). *Technical Report on the Cape Bathurst, Bluenose-West, and Bluenose-East Barren-ground Caribou Herds*. Yelloknife: Department of Environment and Natural Resources, Government of the Northwest Territories.

Diadem Resources (2011). Diadem announces 100% acquisition of diamond rights on the Franklin Project. *MarketWired*. http://www.marketwired.com/press-release/ diadem-announces-100-acquisition-of-diamond-rights-on-the-franklin-project-tsx-v-drl-1568203.htm

Donaldson, S.G., Van Oostdam, J., Tikhonov, C., Feeley, M., Armstrong, B., Ayotte, P., . . . Shearer, R.G. (2010). Environmental contaminants and human health in the Canadian Arctic. *Science of the Total Environment, 408*(22), 5165–5234.

Duhaime, G., & Bernard, N. (2002). Regional and circumpolar conditions for food security. In G. Duhaime (Ed.), *Sustainable Food Security in the Arctic: State of Knowledge* (pp. 227–238). Edmonton: CCI Press, University of Alberta.

Duhaime, G., & Caron, A. (2006). The economy of the circumpolar Arctic. In S. Glomsrød & L. Aslaksen (Eds.), *The Economy of the North* (pp. 17–23). Oslo-Kongsvinger: Statistics Norway.

Duhaime, G., & Caron, A. (2009). Economic and social conditions of Arctic regions. In S. Glomsrød & I. Aslaksen, (Eds.), *The Economy of the North 2008* (pp. 11–23). Oslo-Kongsvinger: Statistics Norway. http://www.ssb.no/a/english/publikasjoner/pdf/ sa112_en/kap2.pdf.

Duhaime, G., Dewailly, E., Halley, P., Furgal, C., Bernard, N., & Godmaire, A. (2008). Sustainable food security in the Canadian Arctic: An integrated synthesis and action plan. In G. Duhaime & N. Bernard (Eds.), *Arctic Food Security* (pp. 73–104). Edmonton: Canadian Circumpolar Institute Press.

Environment and Natural Resources. (2010). *Our Wildlife – Bluenose-West Herd*. Yellowknife: Department of Environment and Natural Resources, Government of the Northwest Territories.

Environment and Natural Resources. (2011). *Caribou Forever – Our Heritage, Our Responsibility: A Barren-Ground Caribou Management Strategy for the Northwest Territories, 2011–2015*. Yellowknife: Department of Environment and Natural Resources, Government of the Northwest Territories.

Food and Agriculture Organization. (2015). *The State of Food Insecurity in the World 2015*. Rome: United Nations.

Ford, J.D., Pearce, T., Duerden, F., Furgal, C., & Smit, B. (2010). Climate change policy responses for Canada's Inuit population: The importance of and opportunities for adaptation. *Global Environmental Change, 20*(1), 177–91.

Inuvialuit Joint Secretariat (2003). *Inuvialuit Harvest Study: Data and Methods Report 1988–1997*. Inuvik: Inuvialuit Joint Secretariat.

Keeping, J. (1999). The legal and constitutional basis for benefits agreements: A summary. *Northern Perspectives, 25*(4). Retrieved from http://www.carc.org/pubs/ v25no4/3.htm

Kerkvliet, J., & Nebesky, W. (1997). Whaling and wages on Alaska's North Slope: A time allocation approach to natural resource use. *Economic Development and Cultural Change, 45*(3), 651–665.

Kruse, J. (1991). Alaska Inupiat subsistence and wage employment patterns: Understanding individual choice. *Human Organization, 50*(4), 317–326.

Kruse, J., Poppel, B., Abryutina, L., Duhaime, G., Martin, S., Poppel, M., . . . Hanna, V. (2008). Survey of Living Conditions in the Arctic (SLiCA). In V. Møller, D. Huschka, & A.C. Michalos (Eds.), *Barometers of Quality of Life around the Globe: How Are We Doing?* (pp. 107–134). Netherlands: Springer.

Lambden, J., Receveur, O., & Kuhnlein, H.V. (2007). Traditional food attributes must be included in studies of food security in the Canadian Arctic. *International Journal of Circumpolar Health, 66*(4), 308–319.

Langdon, S.J. (1986). Contradictions in Alaskan Native economy and society. In S.J. Langdon (Ed.), *Contemporary Alaskan Native Economics* (pp. 29–46). Lanham, MD: University Press of America.

Larsen, J.N., & Huskey, L. (2015). The Arctic economy in a global context. In B. Evengård, J.N. Larsen, & Ø. Ravna (Eds.), *The New Arctic* (pp. 159–174). Berlin: Springer.

Mackenzie Gas Project (2004). Socio-economic effects summary. In *Environmental Impact Statement for the Mackenzie Gas Project,* Volume 1, *Overview and Impact Summary* (Section 6). http://www.mackenziegasproject.com/theProject/regulatoryProcess/applicationSubmission/Documents/MGP_EIS_Vol1_Section_6_S.pdf.

Mackenzie Gas Project (2005). *Environmental Impact Statement for the Mackenzie Gas Project.* Volume 6, Part C, *Socio-Economic Impact Assessment: Paulatuk Community Report.* http://www.mackenziegasproject.com/theProject/regulatoryProcess/applicationSubmission/Documents/Vol_6C-Paulatuk_SEIA.pdf.

Mares, T.M., & Peña, D.G. (2011). Environmental and food justice. In A.H. Alkon & J. Agyeman (Eds.), *Cultivating Food Justice: Race, Class, and Sustainability* (pp. 197–220). Cambridge, MA: MIT Press.

McDonnell, S. (1983). *Community resistance, land use and wage labour in Paulatuk, N.W.T.* (MA thesis). University of British Columbia, Vancouver.

Mintz, S.W., & Du Bois, C.M. (2002). The anthropology of food and eating. *Annual Review of Anthropology, 31*(1), 99–119.

Nadasdy, P. (2003). Reevaluating the co-management success story. *Arctic, 56*(4), 367–380.

Nagy, J.A. (2009). *Population Estimates for the Cape Bathurst and Bluenose-West Barren-Ground Caribou Herds Using Post-calving Photography.* Yellowknife: Department of Resources, Wildlife and Economic Development, Government of the Northwest Territories. http://wrrb.ca/sites/default/files/Nagy%202009%20Pop%20Est%20Post%20Calving%20CB%20BW%20DRAFT.pdf

Nuttall, M., Berkes, F., Forbes, B., Kofinas, G., Vlassova, T., & Wenzel, G. (2005). Hunting, herding, fishing and gathering: Indigenous peoples and renewable resource use in the Arctic. In C. Symon, L. Arris, & B. Heal (Eds.), *Arctic Climate Impact Assessment* (pp. 649–690). New York: Cambridge University Press.

NWT Bureau of Statistics (2007). *Paulatuk: Statistical Profile.* Yellowknife: Government of the Northwest Territories. Accessed 2008. http://www.stats.gov.nt.ca/Infrastructure/CommSheets/Paulatuk.html.

NWT Bureau of Statistics (2010). *Paulatuk: Statistical Profile.* Yellowknife: Government of the Northwest Territories. http://www.statsnwt.ca/community-data/infrastructure/Paulatuk.html.

NWT Bureau of Statistics (2012). *Community Data.* Yellowknife: Government of the Northwest Territories. http://www.statsnwt.ca/community-data/infrastructure/Paulatuk.html.

Paci, C.J., Dickson, C., Nickels, S., Chan, L., & Furgal, C. (2004). Food security of northern Indigenous peoples in a time of uncertainty. Paper presented at the Third Northern Research Forum, Yellowknife.

Peloquin, C., & Berkes, F. (2009). Local knowledge, subsistence harvests, and social-ecological complexity in James Bay. *Human Ecology*, *37*(5), 533–545.

Pottier, J. (1999). *Anthropology of Food: The Social Dynamics of Food Security*. Malden, MA: Blackwell.

Power, E.M. (2008). Conceptualizing food security for Aboriginal people in Canada. *Canadian Journal of Public Health*, *99*(2), 95–97.

Sandlos, J. (2007). *Hunters on the Margin: Native People and Wildlife Conservation in the Northwest Territories*. Vancouver: UBC Press.

Southcott, C. (Ed.). (2015). *Northern Communities Working Together: The Social Economy of Canada's North*. Toronto: University of Toronto Press.

Stabler, J. (1990). A utility analysis of activity patterns of Native males in the Northwest Territories. *Economic Development and Cultural Change*, *39*(1), 47–60.

Todd, Z. (2010). *Food security in Paulatuk, NT: Opportunities and challenges of a changing community economy* (MSc thesis). University of Alberta, Edmonton.

Usher, P.J. (1971). *The Bankslanders: Economy and Ecology of a Frontier Trapping Community* (Vol. 2). Ottawa: Department of Indian Affairs and Northern Development.

Usher, P.J. (2002). Inuvialuit use of the Beaufort Sea and its resources, 1960–2000. *Arctic*, *55*(2), 18–28.

Usher, P.J., Duhaime, G., & Searles, E. (2003). The household as an economic unit in Arctic Aboriginal communities, and its measurement by means of a comprehensive survey. *Social Indicators Research*, *61*(2), 175–202. https://doi.org/10.1023/a:1021344707027

Willows, N.D. (2005). Determinants of healthy eating in Aboriginal peoples in Canada. *Canadian Journal of Public Health*, *96*(3), 32–36.

10

Caribou and the Politics of Sharing

Tobi Jeans Maracle, Glenna Tetlichi, Norma Kassi,
and David Natcher

For centuries, caribou have served as the basis of the economy of the Vuntut Gwitchin people of the Yukon Territory. Providing food, clothing, and many other material needs, caribou have also been essential to the foundation of Vuntut Gwitchin culture because of their symbolic value. Over time, the Vuntut Gwitchin have learned to adapt to the temporal and spatial variability of caribou migrations, with caribou being plentiful in some years and less so in others. Under these conditions, food sharing has helped to minimize the impacts affecting an individual, household, or community when, for various reasons, there has been a lack of access to caribou.

Today, the sharing of caribou remains an important and widely practised tradition among the Vuntut Gwitchin. The basic purpose of sharing caribou has generally remained the same, namely to maximize the overall well-being of the Gwich'in people, while uniting families and communities on economic, social, and political grounds. As in the past, the sharing of caribou not only facilitates the distribution of an economic and nutritional resource but also affirms personal relationships and the social networks that support them.

Over the past century, and particularly since the 1990s, the ability of the Vuntut Gwitchin to share caribou, as well as other country foods, has been progressively restricted due to the imposition and enforcement of the US-Canada border that now separates the Vuntut Gwitchin from their families and friends residing in Alaska. Based on interviews conducted in Old Crow and Fort Yukon in 2010, together with document analysis, this chapter examines how, over the past century, the international border that transects the Gwich'in territory has come to affect the Vuntut Gwitchin and the politics that now surround the sharing of caribou.

BACKGROUND

Old Crow is the most northerly community in the Yukon Territory and the only community in the Yukon without road or marine access. Located at the confluence of the Crow and Porcupine Rivers, Old Crow is 800 kilometres north of Whitehorse and 90 kilometres east of the Alaska border. Today, approximately 300 people reside in Old Crow, 270 of whom are Vuntut Gwitchin. The Vuntut Gwitchin, or "people of the lakes," are part of the Gwich'in people, whose traditional territory extends across western Alaska, through the Yukon, and into the Northwest Territories.

As described by Morlan (1973) and others (Balikci, 1963; Leechman, 1954; Osgood, 1934, 1936a, 1936b;), the Vuntut Gwitchin traditionally followed a pattern of seasonal mobility. In the spring, the Vuntut Gwitchin would concentrate their efforts on hunting migrating caribou at known river crossings along the Porcupine River (Leechman, 1954; Morlan, 1973) between the Bell River to the east and the Coleen River to the west in Alaska (McKennan, 1965). In the late spring, muskrat and bird hunting took place throughout the Old Crow Flats (Leechman, 1954; Morlan, 1972). In the summer, the Vuntut Gwitchin would disperse into camps located along tributaries of the Porcupine and Old Crow Rivers, where fish traps were set for salmon and other species of fish (Leechman, 1954). Other summer activities included egg and berry gathering, rabbit snaring, and in the late summer, capturing molting birds (Leechman, 1954; Morlan, 1973). In the fall, the Vuntut Gwitchin would move to the northern edge of the Old Crow Flats, where they would construct and/or mend caribou fences and surrounds that had been used successfully for generations to trap and kill large numbers of migrating caribou (Leechman, 1954).

Due primarily to geography, the Vuntut Gwitchin remained relatively isolated from European encroachment until the beginning of the nineteenth century (VanStone, 1974). Vuntut territory was said to be one of the most remote fur trade destinations in Canada, having only three reasonable routes of access: up the Yukon River through Alaska, south on the Mackenzie River from eastern Canada, or by ship through the Arctic Ocean and then south across difficult northern terrain (Leechman, 1954). This remoteness, however, did not restrict Vuntut trade with

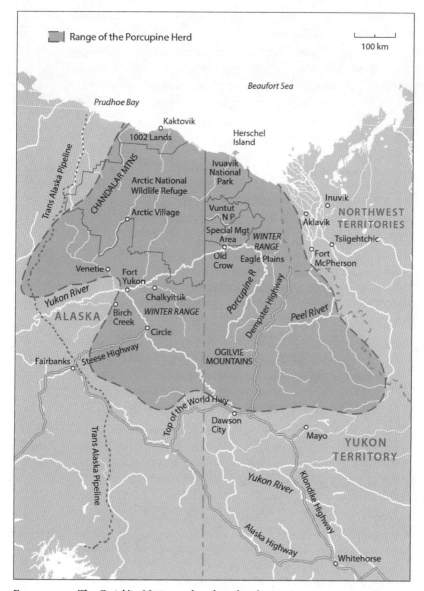

FIGURE 10.1   The Gwich'in Nation and caribou distribution

neighbouring peoples. In fact, the Vuntut were reported to have travelled considerable distances to trade with other Aboriginal groups and
were viewed as astute middlemen by European traders (Hadleigh-West,
1963). Most notable among Vuntut traders were Olti, who would travel
to Hershel Island to trade caribou skins with Inuit; Khach'oodaayu, who

travelled throughout the Mackenzie River system to Fort Simpson; and others who would regularly travel as far as Barrow, Alaska, to trade for Russian goods (Vuntut Gwitchin First Nation & Smith, 2009).

The latter half of the nineteenth century ushered in greater European contact and influence. With increased involvement in the fur trade, Gwich'in settlement patterns began to change as permanent log homes were built near trading posts, replacing to some extent seasonal trapping and fishing camps (VanStone, 1974). Despite becoming more centralized, trade with neighbouring nations continued as it had for centuries, particularly with Inuit of Hershel Island and other Gwich'in residing at Potato Hill, Arctic Village, Rampart House, and Blue Fish. During this time, the Vuntut Gwitchin maintained their independence and continued close personal contact with neighbouring nations through trade and exchange (Leechman, 1954).

US PURCHASE OF ALASKA

On March 30, 1867, the United States Senate approved the $7.2 million purchase of Alaska from Russia. Despite this territorial demarcation, Gwich'in travel to and from Alaska remained largely unaffected. As noted by research participants interviewed in 2010, such as Paul Herbert of Fort Yukon, Alaska, "Oh yeah, many didn't even know the border was there. People from Fort Yukon, where I come from, would go ratting on the Crow Flats [Yukon]." Donald Frost recalled similar freedoms: "There was no real border at that time. At least for the people like us, Gwich'in people could go wherever you want."

Yet with the purchase of Alaska by the United States, the Gwich'in territory had nonetheless come under the administrative authority of two governments – the United States and Canada. With the 141st meridian being officially declared the international boundary separating Alaska and the Yukon, new restrictions came into effect, and by 1912 the Gwich'in had begun to experience the effects of having an international border transect Gwich'in territory. John Joe Kyikavichik said, "They [Gwich'in] used to declare their fur at the border, but they would really do what they wanted. But after 1912, that all changed. If they went across the border and were caught hunting, they would have to pay a large sum of money." Dick Nukon of the Han First Nation explained, "Border stopped people. Once the border line was established, everything changed." In subsequent decades, hunting, trapping, and even travelling in a large portion of the Vuntut Gwitchin's traditional territory were increasingly treated as criminal offences. John Joe Kyikavichik said, "They [government] made a law for everything, caribou, fish, rabbit, porcupine. Even if you had relatives on other side and this side, you still could not hunt on either side."

Stories of the border and dictatorial enforcement officers continue to be told today. The social life of these stories (Cruikshank, 1998) contributes to the fear of country foods being confiscated when travelling to Alaska or even being put into jail for bringing country foods across the border. The stories that circulate are the result of past incidents involving the Royal Canadian Mounted Police (RCMP), US Customs, and Wildlife Enforcement officers. For example, stories of these officers stationed at Moosehide, Alaska, from over seventy years ago are fresh in the memories of elders. Dick Nukon, who was born around Eagle, Alaska, and grew up in various places along the Yukon River, recalled an encounter his father had with an officer: "He [father] went to Dawson and got some food and goods. Game warden was there. My dad talked to the game warden, it was bad ... He didn't let us bring food back into Alaska. Sam White, the game warden on the Alaska side, was very strict." Paul Herbert shared a story of a hunter who shot a duck on the Alaska side of the border while travelling to Old Crow: "The RCMP stopped him and he had a duck in the boat and he got cited for it. He killed it in America, but he had the duck lying in the boat and he got cited for it just on the other side of the border."

The most prevalent story told was that of Trimble Gilbert of Arctic Village. Trimble Gilbert is a fiddler who came to Old Crow for a fiddle festival. As part of a raffle, he won a pair of beaded slippers, made of caribou hide. Upon arrival at the customs office in Fairbanks, Alaska, he declared his gift and customs officials promptly confiscated the slippers. Three members of the Vuntut Gwitchin told this story in great detail, and several others in Old Crow and Fort Yukon acknowledged the incident. From this single event, the fear of taking any traditional items across the border has been created and continues to permeate border discussions to this day. It was unclear whether the slippers were ever returned to Trimble Gilbert, but the rumour is that they were sent to Anchorage, where they were likely kept by the customs supervisor.

Stories were also told of customs agents stationed in Fort Yukon who were known to burn food and other personal items on the airport runway if attempts were made to bring undeclared items across the border.

Although there has not been a border agent in Fort Yukon since the 1990s, the stories are still told today and act as significant deterrents for those considering bringing food or other gifts to Fort Yukon. As Esau Nukon explained, "There could be the wrong person there that don't understand, ya know. The policy of traditional food sharing, you can get pulled over and get canned or evicted for bringing wildlife over because of misunderstanding."

Land Claims and the Right to Share

In 1992 the Yukon Umbrella Final Agreement was signed. One year later, the Vuntut Gwitchin signed their own agreement, and in 1995 the Vuntut Gwitchin First Nation Final Agreement came into effect. With the signing of the agreement in 1993, Vuntut rights to traditional lands and resources were affirmed and protected, including a guaranteed right to allowable harvest levels of subsistence foods, as well as harvesting rights throughout their territory, the boundaries of which were defined by negotiators of the agreement and are not necessarily reflective of the full extent of the Vuntut Gwitchin traditional territory. The agreement also guarantees the Vuntut

> the right to give, trade, barter or sell among themselves and with benefi-
> ciaries of adjacent Transboundary Agreements *in Canada* all Edible Fish or
> Wildlife Products harvested by them pursuant to 16.4.2, or limited pursu-
> ant to a Basic Needs Level allocation or pursuant to a basic needs alloca-
> tion of Salmon, in order to maintain traditional sharing among Yukon
> Indian People and with beneficiaries of adjacent Transboundary Agree-
> ments for domestic purposes but not for commercial purposes. (16.4.4,
> italics added)

These transboundary agreements were developed to clarify subsist-
ence activities and harvesting rights in areas where the Vuntut Gwitchin traditional territory overlapped with land claims or traditional territor-
ies of neighbouring Aboriginal governments. However, the Vuntut final agreement does not guarantee the right of the Vuntut to transport and share caribou or other country foods with family and friends in Alaska. Rather, the Vuntut Gwitchin are required to secure an export permit prior to transporting any food or other goods across the Alaska-Yukon border. The Canadian Customs and Revenue Agency is responsible for

the administration and the issuing of permits under the authority of the Games Export Act of 1985, which regulates the transportation of wildlife products across borders with Alaska, British Columbia, and the Northwest Territories. Following the signing of the final agreement, the federal government did agree to modify the Games Export Act so that the Vuntut could transport goods for traditional and noncommercial purposes across all three borders. Unfortunately, such modifications in the necessary legislation have yet to be made, so today federal legislation still requires the Vuntut Gwitchin to complete an export permit for approval prior to transporting caribou to Alaska.

Failing to secure an export permit can result in a number of responses from government, ranging from turning a blind eye to laying actual charges. However, in cases where caribou are being sent to family and friends in Alaska, a Yukon government representative noted, "We would likely talk to the persons and explain the rules and the reasons for them to get the permit. Maybe a written warning ... We would not lay a charge until we had first presented the case to our headquarters and to the Crown prosecutor for consideration." Referrals by conservation officers either to their headquarters within the Yukon Department of the Environment or to the Crown prosecutor would generally occur only when the amount being shared exceeded what would normally be considered by government as subsistence use. Officials admit that there is considerable ambiguity in the process, with one Yukon government representative noting, "Some wildlife they [Gwich'in] can freely have, give away, send away, just like they always have, and in other situations, they can't." In an effort to clarify the process, the Yukon government offers the following guide, as explained by another representative: "If you harvested it to eat it, in your traditional territory, then it is subsistence, and everything is okay. If you harvested the animal for any reason other than to eat it, all rules apply, and you should check with us before you do anything because likely permits, licence, and other forms of documentation are required."

Acquiring the necessary permit can however be problematic. Aside from the general complexity of the forms, the issuing agency for the Vuntut Gwitchin is the RCMP detachment in Old Crow. Although the relationship between the RCMP and the community is generally positive, there can be some reluctance among Vuntut Gwitchin to approach RCMP staff about acquiring the necessary permits and approvals. The apprehension of being questioned about other activities that are perhaps unrelated and perfectly legal is enough to dissuade some community

members from completing the necessary paperwork. With the responsibility of issuing export permits in the hands of the RCMP detachment,
there is also the perception that severe repercussions can result if a permit
is not acquired.

During interviews, it became clear that many Vuntut Gwitchin were
uncertain and to some extent fearful of the regulations surrounding the
sharing of food with relatives in Alaska. In the case of US Customs and
Border Protection, a staff member admitted that high staff turnover often
results in inaccurate and inconsistent information being shared: "The reason [for the confusion] is that the RCMP are doing them for us; they are
seldom familiar with wildlife laws or the harvesting chapters of the land
claim agreements, [and] they are seldom stationed there long enough
to get familiar with it." In a study conducted in western Alaska, it was
found that the regulations surrounding customary trade are so poorly
understood by community members that uncertainty concerning rules
and regulations likely affects the extent to which customary trade and
barter now occur (Magdanz, Utermohle, & Wolfe, 2002). Uncertainty
over cross-border travel also includes fear of having food and hunting
equipment confiscated by border authorities if there is suspicion of illegal
hunting in Alaska. For most people in Old Crow, this is a risk that simply
cannot be taken. As Billy Bruce said, "I make [a] few trips down to Fort
Yukon with boat, but I was scared to take meat from here down over the
border. I was scared to take it down for my relatives."

Despite concerns over confiscations or even being charged criminally, some Vuntut Gwitchin remain steadfast about the need and right to
share food with friends and family in Alaska. One participant explained,
"Whether or not we dot all the i's and cross the t's on the forms is another
matter. We do manage to get some of those trade and barter goods to
our friends and families across the Gwich'in Nation." Another said, "If
you open the doors to the police today, they come check your boat, or
the border patrol to come check your boat, then they have to apply the
restrictions and the laws. So just go about it and do your business and to
heck with all this. This is the way we have always done it."

The events of September 11, 2001, and the heightened sense of border
enforcement seem to have exacerbated an already serious problem. As
Dennis Frost made clear, security at the border, however gradual and
inconsistently applied, has now intensified:

You know, one set of laws for the United States border crossing applied
to everybody, it never used to be like that. Even getting on a plane from

here [Old Crow], border officers [used to] look at you and see you're First Nation, and they just let you go sometimes. They hardly ever bother with asking too serious business about your passports and stuff. But now, they treat everybody the same. It's not like it used to be, things are different.

"Hard Time Is Coming"

Throughout this project, community members in Old Crow often spoke of hard times that were coming, describing what the land was telling them and the need to be prepared when hard times come. David Lord emphasized that being prepared involves sharing food with those in need: "The hard time is coming and we gotta learn to share more of what we have ... That's the way they used to do it long ago when, and I don't see anything wrong with it. You gotta let the government know that this is our way of life. And hope they understand it." Esau Nukon added, "Today we face the hardship of global warming and climate change, and how do we secure and prevent our young generation for the future hardship, how do we do that, and how do we share with our neighbours, and how do you get rid of that cross-border issue? It's a big topic ... hardship is coming."

Hard times were also linked to declining food sources and the effects of not having access to country foods. As Donald Frost said, "Sometimes it's hard to get caribou ... People come up from Fort Yukon ... anyway 'cause we always get caribou for them ... But not all the time. Some years it get late in the fall, and they go back without caribou because it's late and the ice start running and ... don't get anything." Billy Bruce explained, "'Cause they got certain time to hunt moose in Alaska and that closure is affecting them lots, 'cause that's the only source of food they get, moose, and they don't get caribou." Darius Elias, a member of the Yukon Legislative Assembly, expressed concerns over the inability of his friends and family in Old Crow to harvest country foods and spoke of his personal experience with declining food sources: "I had the opportunity to fish in the Stikine River for sockeye salmon. This year was a terrible year. I got ten fish, and usually I bring home thirty to share with people. But I am still going to share those ten fish, even if it is a half, just so they can taste it."

Coupled with the decline or limited availability of country foods are the high cost of subsistence harvesting and related travel. Stories arose during interviews about flights once being chartered from Arctic Village, Alaska, to give caribou to relatives in Old Crow in times of need. Old Crow did the same when other communities were in need. Florence Netro said, "One year I remember when I was younger, we had no caribou in Old Crow. Caribou didn't come and no one had meat, so they sent a whole bunch of meat from Inuvik, yeah, oh just big planeload. The whole town got meat. It was good ... then later ... we sent meat over there." Given the high cost of fuel and plane charters, air travel is unfortunately an option that no longer exists. As Darius Elias remembered, "Years ago we used to transport planeloads of caribou when a village or community wasn't able to harvest, and it went both ways, too. Some planes would land here with caribou, and sometimes we would ship planes of caribou meat other places. That is pretty tough to do nowadays."

For some, sharing has either been impeded or stopped completely due to the high cost of fuel and travel. Teresa Frost explained, "You can only take so much with you, a little box, and then transport, and they have to look through it, and it is not like you can charter right to Fort Yukon. You can't. You have to go to Fairbanks, then to Fort Yukon. So still it's more expensive." Brandon Kyikavichik said, "We are at the mercy of things these days. In the old days, we were not at the mercy of anything. We just had our own bodies and that's all we needed. Now we are at the mercy of gas and money, machines and things like that, so we can't do what we used to do long ago."

Many Gwich'in of Fort Yukon who once regularly travelled to Old Crow now forgo those trips due to the chance they will not be able to bring food home. If a trip to Old Crow has the potential to yield no food, gas money is better spent on excursions closer to home, with the resulting harvest shared with family and friends in Fort Yukon. Although a trip to Old Crow, Fort Yukon, or Arctic Village is about harvesting and exchanging country foods, it is also about maintaining cultural traditions of sharing. However, due to the financial costs involved in making such trips, together with the risk of having food confiscated at the border, these social and cultural exchanges have become increasingly limited.

BUILDING FOOD-SHARING NETWORKS ACROSS BORDERS

We should not feel threatened. It is our tradition to trade, you know. We should not be afraid to go across that border and shake hand with a Gwich'in person you know for a long time and then cross border without fear. We should have that freedom. We had that freedom long time ago ... but as soon as that border came, there was restriction and some people paid penalty for it because of regulations. But if we can work together, our young generation will not lose their identity, not lose their family tree, who they are related to. I want to see that continue.

– Vuntut Gwitchin elder

The above appeal mirrors the concerns expressed by Gwich'in on both sides of the border. Since its establishment in the early 1900s, the border has not only come to restrict physical access to Gwich'in territory but also affected the social, cultural, economic, and political ties that have long united the Gwich'in people through the simple act of sharing food. Yet the significance of sharing food is so deeply entrenched within Gwich'in culture that its importance cannot be minimized. One can turn to Gwich'in oral history to appreciate the importance of food sharing, such as the story of the man who became a *nanaa'in* (bushman) by "violating the social obligation to share food" (as told by Moses Tizya in Vuntut Gwitchin First Nation & Smith, 2009, p. 49) or the story of Old Woman, which describes the virtues of sharing food. Roger Kyikavichik pointed out that the barrier impeding food sharing from occurring "carries with it a lot of pain" for the Vuntut Gwitchin.

With the settlement of the Vuntut Gwitchin First Nation Final Agreement, it was hoped that the issue of territorial access and the right to share caribou with family and friends in Alaska would be secured. However, because no explicit statement in the agreement specifies that import and export permits for harvested wildlife are not required, and having yet to establish their own policies for transboundary trade, the Vuntut Gwitchin are required to secure the appropriate permit prior to the transport of country foods. In cases where the necessary permit is not received, the Vuntut Gwitchin can now find themselves in violation of Subsection 6(2) of the Yukon Act of 2002, Sections 6(3) and 7(1) of the Wild Animal and Plant Protection and Regulation of International and Interprovincial Trade Act of 1992, as well as Subsection 105(1) of the Yukon Territory Wildlife Act of 2002. The magnitude of this legislative bureaucracy is made that much more complex by having the traditional territory of the Vuntut Gwitchin administered by one territorial government (Yukon), one state government (Alaska), two federal governments (the United States and Canada), and two comprehensive land claims (the Alaska Native Claims Settlement Act of 1971 and the Vuntut Gwitchin First Nation Final Agreement of 1993). Arguably, the Vuntut Gwitchin now find themselves with more territorial restrictions upon their lives and livelihoods than any other Aboriginal group in North America.

Compounding the legalities that now influence Gwich'in food sharing are the rising costs associated with harvesting and travel between communities to share. The chance of being turned back due to missing or incomplete paperwork, having food confiscated, or even being charged criminally has in many cases deterred families and friends from making trips across the border. For some Gwich'in, this risk has impeded food sharing; for others, it has stopped food sharing all together. Yet a decline in food sources has created an even greater need for sharing food, as warned by those Vuntut elders who agreed that "hard time is coming." The sharing of food will be critical to the survival of the Gwich'in when those hard times come. Although more than thirty other First Nation and Native American tribes are affected by an arbitrary border, or "medicine line," that separates Canada and the United States (O'Brien, 1984), the Vuntut Gwitchin are unique in their degree of isolation and their continued reliance on country foods, particularly in their dependence on migratory populations of caribou. Overcoming the challenges associated with the border will require a concerted effort not only by the Vuntut Gwitchin but by all Gwich'in. One participant said, "Our strength comes from our nation, not individual villages. The only way it will work is to make a nation across two nations, one people, one voice. That's the only way it will work. We belong to the Gwich'in Nation, and it must be recognized by both Canada and the US – all governments."

References

Balikci, A. (1963). *Vunta Kutchin Social Change: A Study of the People of Old Crow, Yukon Territory*. Ottawa: Northern Co-ordination and Research Centre, Department of Northern Affairs and National Resources.

Cruikshank, J. (1998). *The Social Life of Stories: Narrative and Knowledge in the Yukon Territory*. Lincoln: University of Nebraska Press.

Hadleigh-West, F. (1963). *The Netsi Kutchin: An essay in human ecology* (PhD diss.). Louisiana State University, Baton Rouge.

Leechman, D. (1954). *Vanta Kutchin*. Ottawa: National Museum of Canada.

Magdanz, J.S., Utermohle, C.J., & Wolfe, R.J. (2002). *The Production and Distribution of Wild Food in Wales and Deering, Alaska*. Juneau: Division of Subsistence, Alaska Department of Fish and Game. http://www.subsistence.adfg.state.ak.us/techpap/tp259.pdf

McKennan, R.A. (1965). *The Chandalar Kutchin*. Calgary: Arctic Institute of North America.

Morlan, R.E. (1972). *NbVk-1: An Historic Fishing Camp in Old Crow Flats, Northern Yukon Territory*. Ottawa: National Museum of Man.

Morlan, R.E. (1973). *The Later Prehistory of the Middle Porcupine Drainage, Northern Yukon Territory*. Ottawa: National Museum of Man.

O'Brien, S. (1984). The medicine line: A border dividing tribal sovereignty, economies and families. *Fordham Law Review, 53*(2), 471–485.

Osgood, C. (1934). Kutchin tribal distribution and synonmy. *American Anthropologist, 36*(2), 168–179.

Osgood, C. (1936a). *Contributions to the Ethnography of the Kutchin*. New Haven, CT: Yale University Press.

Osgood, C. (1936b). *The Distribution of the Northern Athapaskan Indians*. New Haven, CT; Yale University Press.

Vuntut Gwitchin First Nation & Smith S. (2009). *People of the Lakes: Stories of Our Van Tat Gwich'in Elders/Googwandak Nakhwach'ànjòo Van Tat Gwich'in*. Edmonton: University of Alberta Press.

VanStone, J.W. (1974). *Athapaskan Adaptations: Hunters and Fishermen of the Subarctic Forests*. Chicago: Aldine.

PART 4
Governance and Management

11

Recollections of Caribou Use and Management

Robert Charlie

M y earliest recollections of caribou were as a young child travelling somewhere in the Richardson Mountains near the community of Fort McPherson. It was during the winter and I was snuggled in the toboggan, but I was able to see as we crossed the small mountain creeks. As we travelled, the men would point up into the mountains, and you could see the caribou up there against the skyline. I can remember the excitement of the family groups that were travelling to their winter hunting grounds in order to hunt and prepare the caribou for use later in the spring and summer, when meat was scarce.

After travelling for a while, we broke out into an area that was traditionally used to establish the winter camp, and this is where the family groups would spend part of the winter. After camp was set up and everyone was settled, the men would usually gather together to plan for the next day's caribou hunt. Early the next morning before sunrise, the men who were to be part of the hunting party would prepare, with some proceeding ahead on snowshoes to ready the trail for those following behind with dog teams. The men usually had a good idea where the caribou would be, having hunted in this area before and possessing knowledge that was passed down from generation to generation. While the men were out hunting, those who were left behind would prepare the camps by fixing up their tent locations, hauling wood, getting spruce boughs for the tents, and setting up poles for drying the meat.

Sometimes you could hear the men shooting if the caribou were close by, but other times you would have no idea whether the men were successful or not until they returned well after dark. They would be able to bring back only what they were able to pack. Usually, the hunters would bring back the caribou delicacies that are still enjoyed to this day. These parts were cooked up as soon as they were brought home. The following day, the men would use dog teams to haul meat from the hunt location. The women would then begin to prepare to dry the meat and make bone grease. They would utilize all the parts of the animal, including the hide, leg bones, and all the stomach parts. Children also had their chores to do but were allowed to explore around the camp area. In the evenings, they would observe as their parents worked with the caribou meat. This was how the use of the caribou was passed on from generation to generation.

The hunters who went out hunting planned the day's activities. There was usually an elder who gave directions to the young men on the strategy that they would use when they came upon a group of caribou. Each hunter was given direction on his role within the larger hunting party, and as a result they were for the most part successful in their hunt. All the hunters shared in the harvest and worked together to help each other if they ran into difficulties.

Not all hunts were successful, as sometimes you would travel into the mountains to find that the caribou had passed and gone to another area where they may not have gone for a period of time. From the community of Fort McPherson, there are certain creeks and rivers that were used to travel into the mountains. Stoney Creek and the Rat River, Vittrekwa River, Road River, Trail River, and Caribou River all lead into the Richardson Mountains. These were the access routes that hunters with dog teams traditionally used. It was not unusual for hunters to travel using these routes only to find no caribou. It was not unusual for hunters to be away for at least a month if using the Caribou River to get into the mountains. This part of the mountains is the farthest location from Fort McPherson and could take a week to ten days to reach. For the most part, hunters could be away as short as a week or as long as a month. During these hunting trips, they would have to be totally self-sufficient, hauling their tents, stoves, and food for themselves and their dog teams. They usually travelled in groups, as it was easier, and they would share in bringing different items to camp. If they did harvest caribou, they would then have to haul the meat all the way back to Fort McPherson or to their camps along the Peel River.

I can remember the Christmas and Easter holidays, when our parents would return to town for the celebrations. There would be lines and lines of dog teams loaded with young children and all the delicacies our parents had prepared for the celebration. They would have dry meat, bone grease, and all the caribou meat ready and prepared to cook. Everything was shared with those who were unable to travel into the mountains in order to hunt or with those who had accepted wage employment and lived in the community.

Fast-forward now to the early 1970s and the opening of the Dempster Highway, which changed the lives of the Gwich'in with respect to their hunting practices. Their lives had been changed even as early as the late 1950s when the residential school opened in Fort McPherson. Children were now enrolled in the residential school for ten months of the year. The Gwich'in people, who had normally spent the majority of the time on the land, were now in the community, and some people were now making their living from wage employment. Children attending the residential school did not have the opportunity to learn the ways of the Gwich'in, and as a result the respect and dependency on caribou, other species, and the land declined, although the way of life was maintained by those who still depended on the land. Even in the 1960s, with the introduction of snowmobiles, the people's lives were changed when people switched from dog teams to snowmobiles. As a result, they did not have to spend the summers on the land storing fish for the winter.

The Dempster Highway had a major effect on the hunting practices of the Gwich'in people, who had lived a subsistence lifestyle prior to the residential school opening. An even greater impact was felt with the use of vehicles and snowmobiles for hunting and travelling on the land. The Gwich'in were now using vehicles and snowmobiles to drive into the Richardson Mountains in order to harvest their caribou. There was no longer the planning prior to the hunt, there was no sharing, and the respect for the caribou was no longer practised.

With the signing of the Gwich'in Comprehensive Land Claim Agreement on April 22, 1992, we now have the responsibility of managing the resources in the Gwich'in Settlement Area. These resources included all the subsistence wildlife species on which we depended, as well as migratory birds and fish. The agreement now gave us the responsibility to monitor our own hunting practices in order to ensure the sustainability of the resources on which we depended, including the herds of caribou that we had harvested for thousands of years.

The Gwich'in Renewable Resources Board was established as the body responsible for the management of wildlife, fish, migratory birds, and forests in the Gwich'in Settlement Area. This board was of course created in partnership with government agencies such as Environment and Natural

Resources at the territorial level and the Department of Fisheries and Oceans and the Canadian Wildlife Service at the federal level. In addition to the Gwich'in Renewable Resources Board, the Gwich'in Tribal Council has the responsibility to manage these resources.

Today, there are issues of declining numbers of subsistence species due to natural cycles, climate effects, and harvesting activities. All groups that have responsibilities for resource management work together to plan for the use of these resources and to ensure the long-term sustainability of the resources on which we have traditionally depended. We have worked with other co-management boards along the coast and Mackenzie River on transboundary species such as caribou and fish. This cooperation has been very successful, as all groups have a vested interest in sustaining the resources on which we depend. We have had to put management actions into place, but for the most part, these regulations have been short-term, and we have had compliance from all groups affected.

Because of the changing times, we now have to introduce programs into the schools for those who have not been taught how to use the resources wisely and how to respect them. Community residents still enjoy the subsistence lifestyle but only for certain activities and for short periods. We utilize the traditional knowledge of our elders to impart their wisdom to the youth, just as our elders have done for countless generations, but this teaching now occurs more in a classroom setting than on the land. We also plan community hunts, where we hunt as we have traditionally done, spending actual time on the land. There are still parents who take their children out on the land in order to teach them the ways of the Gwich'in, such as respecting the resources, always taking only what you need, using all the parts of what you harvest, and sharing with others.

12

Ways We Respect Caribou: A Comparison of Rules and Rules-in-Use in the Management of the Porcupine Caribou

Kristine Wray

The management of barren-ground caribou populations has been a focus of study for decades, with critical consideration given to the ways that management does or does not reflect the interests of Aboriginal peoples who have, for thousands of years, depended on caribou for their livelihood. The transition of wildlife management from a system of centralized and top-down decision making to a system of co-management involving multiple stakeholders and operating across many ecological and political boundaries has been a particular preoccupation of academics studying the Porcupine caribou herd of the Northwest Territories, Yukon, and Alaska.

Although much has changed over the past half-century, there are concerns that Aboriginal people's interests, and specifically their traditional knowledge, are being persistently marginalized in a highly bureaucratized co-management system (Kofinas, 2005; Nadasdy, 2003). Building on this body of work, this chapter seeks to understand more about the legitimacy of co-management arrangements at the local level by determining the degree of synergy between the formal rules of the territorial governments and a co-management board and the "rules-in-use" for respecting caribou in operation in the day-to-day life of a local community – Fort McPherson – dependent on caribou for subsistence.

Specifically, the chapter compares the local rules-in-use of the Teetł'it Gwich'in and the formal guidelines and regulations of the Governments of the Yukon and Northwest Territories and the Porcupine Caribou Management Board (PCMB). Rules used by the community and those used by government wildlife managers are compared and contrasted for

similarities and differences in content and theme. Major areas of simi-
larlity include avoiding waste and disturbance of caribou, as well as a
strong focus on hunting safety. Given historical conflicts over caribou
between harvesters, biologists, regional Aboriginal authorities, and territorial governments with respect to the Porcupine caribou as well as other
caribou herds (Kofinas, 2005; Nadasdy, 2003), the intent of this analysis
of rules and rules-in-use is to identify common and uncommon ground.
Although guided by theories on co-management and common-pool
resources, I anticipate that by identifying areas of current and potential
conflict between rules and rules-in-use, the analysis can also serve to
strengthen the co-management process in what is arguably a challenging
and uncertain period of wildlife management in northern Canada.

SETTING

The settlement of northern land claims with Canada has led to new
arrangements for managing lands and resources that respect the rights and
interests of Aboriginal peoples. In many parts of the Northwest Territories,
these management arrangements have often taken the form of co-management
processes in which governments and Aboriginal groups work together to
ensure the sustainability of valued resources, such as barren-ground caribou,
while at the same time ensuring that rights and interests set out in land
claim agreements are protected.

DIFFERENT CULTURES, DIFFERENT RULES

Over the past twenty years, the management arrangements governing the
Porcupine caribou herd have proven to be a useful socio-political case
study for those interested in the theories and practice of co-management.
In addition to learning what is working well, we know that a variety of
institutional, epistemological, and socio-cultural conflicts have suggested
some of the challenges associated with co-management in Arctic ecosystems
(Kofinas, 2005, p. 181). Whereas some conflicts may be surface issues of a
technical nature, other kinds of problems have given rise to claims of
inequitable power relations between Aboriginal and non-Aboriginal people,
with the legitimacy and authority of co-management over "caribou" and
"caribou harvesting" being challenged academically and legally. Differences
between the so-called bureaucratic rules of the state and co-management

boards and those already traditionally in use among Aboriginal people are seen as problematic in other resource management contexts (Berkes, 1989; Pinkerton, 1989). A critical question emerging from this body of work on co-management is "whether informal local authority systems of resource management can sustain their legitimacy while nested within larger more dominant institutional processes" (Kofinas, 2005, p. 181). It is in this context that I explore the fit between the rules-in-use of the Teetł'it Gwich'in and those regulations and guidelines around respecting the Porcupine caribou herd created by the Porcupine Caribou Management Board and the Governments of the Yukon and Northwest Territories. Given the overlap of the herd range into Alaska, the rules set out by the Alaska government are also relevant, but they are not within the scope of this study.

Rules

Rules in natural resource management are expressed in many different ways. Rules set out by governments and also by co-management institutions tend to be highly bureaucratized systems of authority and prescription enforceable by legal instruments such as the Canadian justice system. In the case of the Northwest Territories government's authority over caribou, rules fall under the Wildlife Act R.S.N.W.T. 1988, c.W-4, and the Wildlife Act R.S.Y. 2002, c. 229. In the case of the Yukon, the government minister responsible for the Wildlife Act holds authority for all matters pertaining to wildlife, with the main focus being on licensing, harvesting, and hunting safety, whereas harms to wildlife resulting from habitat loss and degradation or broader issues of human activity are of limited consideration.

Within this context, the authority of the Porcupine Caribou Management Board was established through a negotiated agreement between the federal government, territorial governments, and Aboriginal groups signed in 1985. The PCMB consists of eight board members representing six signatories: the Governments of the Northwest Territories, Yukon, and Canada, the Inuvialuit Game Council, the Gwich'in Tribal Council, and the Council of Yukon First Nations (PCMB, 2004–13). The authority of the PCMB on matters related to caribou and caribou habitat is limited to advising government. Neither does it have access to all the human and financial resources necessary to undertake its mandate, including research, monitoring, and communications. As noted by Kofinas (2005, p. 181), "With no authority of its own, it must cultivate and maintain its legitimacy in the management process both with government agencies as well as with local user communities," such as the Teetł'it Gwich'in, who have their own rules-in-use for respecting caribou.

The reported decline of the Porcupine caribou in the past decade spurred numerous responses on the part of the Government of the Northwest Territories and the Porcupine Caribou Management Board. Although it was well established that there were many factors influencing the health of caribou herds, the emphasis was on creation and enforcement of rules for harvesting, including harvesting by Aboriginal peoples (GNWT, 2011; PCMB, 2004–13; PCMB, 2010). As a result, a caribou management plan was developed by the Government of the Northwest Territories (2011), and a harvest management plan was proposed by the Porcupine Caribou Management Board. Although the caribou management plan does not differentiate between Aboriginal and non-Aboriginal hunters, 90 percent of the people who use the Porcupine caribou herd are Aboriginal (PCMB, 2010, p. 7); thus the plan's main focus on regulating, or restricting, harvesting falls mainly upon Aboriginal people.

Rules-in-Use

The term "rules-in-use" in this chapter is informed by the literature on common-pool resources. The concept is one that has currency in the sociology and political science literatures. Research that focuses on social norms, customary laws, guides, directions, taboos, and the limits defining proper and improper behaviour toward the environment is also relevant to this discussion on rules-in-use and rules for respecting caribou (Guédon, 1994; Kofinas, 1998; Nelson, 1983).

The norms for respecting caribou within Gwich'in culture are grounded in a set of values and an understanding of human-caribou relations that are social-ecological in nature. Rules dictating right and wrong behaviour toward nature, if followed, result in the well-being of both people and nature (Nelson, 1983). If they are not followed, with the result that the balance between the animal and human worlds is not maintained, "repercussions will be dramatic" (Sherry & Vuntut Gwitchin First Nation, 1999, p. 212). Previous research in Fort McPherson suggests that the Teetł'it Gwich'in have a variety of traditional rules related to respectful harvesting as well as to meat sharing, preparation and storage of meat, and waste avoidance (Kofinas, 1998). There are also specific rules related to "respecting the caribou leaders," the first bulls that pass during migration. What is meant by "respect," however, can vary significantly from one community to another and within communities, making it difficult to generalize beyond a single place (Padilla, 2010).

Some anthropologists suggest that rules cannot be shared based on one's experience or taught by people at all but rather come from the process of being or dwelling in the environment (Ingold, 2000). Knowledge or "truth" is not something that can be put into words but resides in the beings of the land itself (Davidson-Hunt & Berkes, 2003, p. 5). In other cases, rules may be expressed as an extensive and detailed story to be shared in a particular kind of social group (e.g., elders and their grandchildren). Some rules may be spoken as if generalizable to all members of a community; whereas these rules seem to be framed as applying to everyone, others may be voiced as "that's the way I do it," not as "that's how one should do it" (Guédon, 1994, p. 49). Thus scholars involved in research elsewhere might similarly argue that rules guiding human-caribou relations cannot easily be reduced to a "reified series of descriptive or normative statements" but are better understood as principles of good relations (ibid., p. 61):

> It was equally difficult to reach some kind of agreement on the details of hunting ritual prescriptions. While my informants more or less agreed most of the time on the principles behind the observances, principles in keeping with and validated by the myths used as references for the animals concerned, *they each had a personal version of taboos and rituals which would be followed on different occasions.* (Ibid., p. 67n11, italics added)

Inasmuch as rules are communicated based on one's experience and through oral history, they are also spoken in shorter form. Such short forms (e.g., respect the caribou) seemingly function as shorthand for transmitting significant bodies of knowledge in ways that ensure their persistence over time. The use of such shorthand may be the only means of ensuring compliance with particular norms without the necessity of communicating all meaning in all instances. Despite there being hundreds, if not thousands, of experiences and observations related to respectful human-caribou relations, arguably only thin slices of observation, experience, and belief can be shared at any given time. These thin slices have been referred to as "rules of thumb," which can cut through the infinite levels of complexity and have the added benefit of being easily remembered (Berkes & Berkes, 2009, p. 7). Rules of thumb are thought to be similar to other cultural symbols, categorizations, or naming practices (e.g., place names) that relay information about landscapes, animals, plants, and other ecological processes (Kritsch, Andre, & Kreps, 1994; Johnson & Hunn, 2010). Although rooted in the past, new rules of thumb are continually being created in response to variation and change in communities and the environment. Guided by this literature, I use the concept of "rules of thumb" or "rules-in-use," drawn from the theory of common-pool resources, as the lens for exploring Gwich'in knowledge about respecting caribou.

LEARNING FROM THE TEETŁ'IT GWICH'IN

This section describes research methodology, research activities, and the major thematic questions of the interview guides used in the research. For guidance on methodology, I consulted that of community-based participatory research (CBPR), an offshoot of participatory action research. CBPR was adapted by Fletcher (2003) to relate to the specific context of research with Aboriginal peoples in northern Canada. CBPR emphasizes involving the community in the research as partners rather than subjects, skill development of community members, and knowledge exchange between the researcher and members of the community (ibid.). There were three partners on this project, the Gwich'in Social and Cultural Institute, the Gwich'in Renewable Resource Board, and the Tetlit Gwich'in Renewable Resource Council. (These three organizations were created by the Gwich'in Comprehensive Land Claim Agreement. The Tetlit Gwich'in Renewable Resource Council is a community-level council and is responsible for local resource management projects related specifically to Fort McPherson; the Gwich'in Social and Cultural Institute is responsible for culture, heritage, and language research and programming; and the Gwich'in Renewable Resource Board conserves and manages resources in the Gwich'in Settlement Area.) They each provided guidance on aspects of the project. Specific issues were clarified within a formal research agreement between the researchers and the partners; these issues included the purpose of the research, scope and methods, obligations to the community, consent, ownership and storage of data, and the roles of the cultural institute and the resource board.

Data collection occurred during the periods of October 10 to November 19, 2007, and February 25 to 27, 2008. Research activities included gathering primary data from elders and harvesters, as well as secondary data from the Government of the Northwest Territories and the Porcupine

Caribou Management Board. With regard to primary data collection, 51 people (11 females and 40 males) were interviewed: 27 harvesters, 19 elders, and 5 others. Interviewee ages ranged between 19 and 71, with the following numbers of people in each age category: 7 aged 19 to 29, 5 aged 30 to 39, 7 aged 40 to 49, 6 aged 50 to 59, 1 aged 60 to 69, and 1 aged 70 to 79. Interviews with elders were semi-directed, with 5 guiding questions, and harvester interviews consisted of a 50-question survey with 37 quantitative questions and 13 qualitative questions. Secondary data collection for the Porcupine Caribou Management Board came from the content of its website (PCMB, 2004–13), attending its annual meeting held September 22 to 24, 2007, and written promotional material such as reports, posters, and pamphlets. For secondary data from the Government of the Northwest Territories, I accessed its Environment and Natural Resources website (GNWT, n.d.), its hunting regulations for the years 2009 to 2010 (GNWT, 2009), and the Wildlife Act R.S.N.W.T. 1988, c.W-4.

The elders' question set focused on acceptable and nonacceptable caribou-hunting practices. Interviews with elders were completed before approaching harvesters. Through these interviews, elders sketched out the ways that life had changed for the community over their lifetimes. These interviews provided an understanding of how people lived, hunted, and related to caribou when elders were young, personal views on how people live and hunt in the present, and a valuable context within which to base the research. Harvester interviews focused in depth on harvesting activities and behaviours, perceptions of caribou health and well-being, and the multiple sources from which hunters get information about caribou. From these interviews, I was able to gain insight into how rules are communicated and whether the mode of communicating rules has changed across the generations.

Teetł'it Gwich'in entries in the tables below are a compilation of elders' and harvesters' responses to particular questions. The questions were, What kind of traditional practices do you think are important to remember in caribou hunting? (asked of harvesters) and, What are traditional practices for respecting caribou? (asked of elders). Also asked were the questions, Have these practices changed since you were young? and How have the Dempster Highway, snowmobiles, and trucks changed how people respect caribou? The Government of the Northwest Territories entries are taken from its hunting regulations for 2009 to 2010 (GNWT, 2009), from the Wildlife Act R.S.N.W.T. 1988, c.W-4, and from discussions with the Fort McPherson renewable resource officer, who represents Environment and Natural Resources. Lastly, the Porcupine Caribou

Management Board entries arose from its printed posters and pamphlets, its website (PCMB, 2004–13), and its *Harvest Management Plan for the Porcupine Caribou Herd in Canada* (PCMB, 2010).

Teetł'it Gwich'in Rules-in-Use for Managing Caribou: How Do Government Rules Compare?

The comparison of the Teetł'it Gwich'in rules-in-use with the rules of the Government of the Northwest Territories and the Porcupine Caribou Management Board (see Tables 12.1 and 12.2) reveal many areas of agreement and few apparent conflicts. Areas of agreement include the importance of not unduly disturbing caribou, not wasting caribou meat during the hunt or in the preparation and storage of meat, and not leaving wounded caribou behind to die unused by the hunters and their families. All three parties agree that during the rutting season the meat of caribou bulls is inedible due to high levels of hormones in the body, and thus all agree that hunting bulls during this time of the year is inadvisable. Lastly, to ensure the safety of hunters, particular practices designed to reduce danger to hunters are agreed upon. There were fewer areas of conflict, and two of the three areas presented – what is considered waste and the feeding of caribou meat to dogs – are becoming less of a conflict as time goes on. The younger hunters tend to utilize less of the caribou, such as the guts, than do the older hunters, who are familiar with how to prepare and use all the parts of the caribou. Younger hunters are thus more inclined to have the same perspective as the territorial government on what is acceptable to leave behind than elders may have. Similarly, feeding caribou meat to dogs is less of a conflict in the present than it was fifty years ago. Since the advent of highway hunting, very few, if any, people use dog teams and thus there are no large teams of dogs to feed. The hunting of female caribou is the only ongoing conflict presented.

The remainder of this discussion considers the way the rules-in-use of the Gwich'in and the rules of the territorial government and the management board are formulated and presented, before moving to broader generalizations about the autonomy granted to individuals with respect to these rules. Gwich'in rules-in-use are presented in short, simple phrases. They refer to an underlying body of cultural information and ways of doing things that is not communicated within the phrases themselves. The cultural context is understood mainly by the Gwich'in community and less so by those external to it. For example, the Gwich'in say, "Don't

TABLE 12.1 Complementary rules or rules-in-use for caribou hunting in the Gwich'in region

	Teetł'it Gwich'in	Government of the Northwest Territories	Porcupine Caribou Management Board
Disturbing caribou	"Don't chase caribou with snowmobiles."	"No one may chase, harass or molest wildlife"(GNWT, 2009, p. 5).	"Reduce stress to caribou ... The use of snowmachines to hunt caribou is the biggest factor in spooking caribou ... Ensure caribou meat is the best quality possible and prevent injury to other caribou" (PCMB, 2010, p. 29).
Waste	"Take what you need, use all that you take, don't waste."	"It is an offence to waste, destroy, abandon, or allow to spoil: the meat of big game, other than bear, wolf or wolverine" (GNWT, 2009, p. 5).	Recommends meat-care methods, proper firearms and ammunition, and sex and age selection (PCMB, 2004–13).
Wounding loss	"Don't leave wounded caribou."	"A person who wounds wildlife shall make every reasonable effort to retrieve it" (R.S.N.W.T. 1988, c. W-4).	"A wounded animal should be immediately shot again to kill it" (PCMB, 2010, p. 29).
Rutting bulls	"Don't shoot rutting bulls."	Does not address the issue.	"Take any bull until October 8 when the rutting season starts ... During the rut until mid November, take small antlered bulls" (PCMB, 2010, p. 30).
Hunting safety	"Don't shoot toward people, watch where you shoot, and never shoot toward the highway. Know who is around when you are shooting."	"No one shall hunt wildlife without due regard for the safety of other people and property. No person shall hunt or discharge a firearm from, or within, a motorized vehicle. In addition, no person shall have in, or on, a vehicle a firearm that has any propellant powder, projectile or cartridge that can be discharged in the breech or firing chamber. No one shall discharge a firearm from, along or across a public road" (GNWT, 2009, p. 5).	"Take the time to notice other hunters or other caribou around your quarry before shooting." (PCMB 2010, p. 28) "Never shoot toward a road or down the travelled portion of a road." (PCMB 2010, p. 28) "Wear blaze orange so other hunters can see you." (PCMB 2010, p. 28)

TABLE 12.2 Conflicting rules or rules-in-use for caribou hunting in the Gwich'in region

	Teetł'it Gwich'in	Government of the Northwest Territories	Porcupine Caribou Management Board
Defining waste	"Bring everything home, can use all parts of the caribou, especially guts and head, shot injured meat can be fed to dogs."	"With regards to ungulates, the following are not considered waste if they are left behind; the head, the legs below the knee joints and the internal organs. Bones, including rib bones, that are stripped of meat may also be left behind. The shot damaged parts of the carcass may also be cut away and left behind" (GNWT, 2009, p. 5).	Recommends meat-care methods, proper firearms and ammunition, and sex and age selection (PCMB, 2004–13). "According to tradition, all parts of the caribou are used; there is no waste. To this day, the skins are used to make traditional clothing from head to toe – from hair pieces to moccasins – and ornamented with beadwork. Furs line mukluks and parkas for warmth and decoration. Bone and antler are fashioned into tools. The caribou heads are roasted over a fire and eaten. For special feasts, a delicacy of head soup is served. Bone marrow is extracted, cooked and eaten. The hooves are jellied and eaten" (PCMB, 2004–13).
Caribou as dog food	"Caribou meat and guts fed to dogs, especially in the past when dog teams were relied on for transportation."	"It is an offence to feed the meat of big game, other than bear, wolf and wolverine, to domestic animals" (GNWT, 2009, p. 5).	Does not address the issue.

| Hunting caribou cows | Multiple perspectives for and against. | "The Gwich'in have the right to harvest all species of wildlife within the settlement area at all seasons of the year subject to limitations which may be proscribed in accordance with this agreement" (Canada & Gwich'in Tribal Council, 1992, p. 44).

The "importance of harvesting bulls, not cows," is one of the messages that the territorial government communicates to the community (Environment and Natural Resources wildlife officer, pers. comm., November 6, 2007). | "We have also requested that all hunters voluntarily avoid hunting female caribou so that the herd's declining population has the best chance to recover" (PCMB, 2004–13).

"Spare the cow – if a hunter chooses one bull instead of a cow each year for 10 years, there will be 23 more caribou in the herd. This isn't enough of a change to reverse the population's downward trend" (PCMB, 2004–13). |

chase caribou with snowmobiles," but it is unclear to a non-Gwich'in person what exactly constitutes "chasing by snowmobile." When considering the seemingly simple phraseology of the Gwich'in rules-in-use, it should also be considered that Gwich'in rules are translations of concepts originally presented and formulated in the Gwich'in language, Dinjii Zhu' Ginjik, and that translation into English may necessitate a simplicity in presentation. Further, there is "interpretive space" within the Gwich'in phrases that reflects the cultural autonomy accorded to Gwich'in hunters to make decisions according to the specifics of the context. Expanding on the earlier example, it is up to a Gwich'in hunter to decide what is an appropriate way to use a snowmobile during caribou hunting. Some things that may influence the decision are distance from the highway, snow and weather conditions, time of day, the number of caribou available, and many other potential factors.

Unlike the Gwich'in rules-in-use, the regulations of the territorial government are very specific and attempt to be very clear in what they are communicating. The specificity of the government regulations can be attributed to the strong interest of government in allowing as little room as possible for interpretation and variation in hunter behaviour. These regulations direct people to do and not to do very specific things. In the regulation against harassing caribou, the Wildlife Act R.S.N.W.T. 1988, c.W-4 attempts to describe in detail what harassment of caribou is: to "persistently or repeatedly chase, weary, harass or molest wildlife without intending to capture or kill it." Unlike the Gwich'in rules-in-use, there is little room within this rule set for autonomy in decision making or for interpretation of rules according to context.

Statements of the Porcupine Caribou Management Board tend to be recommendations rather than rules or regulations in that they suggest hunting behaviours and describe how to achieve them. This approach is taken because the PCMB is an advisory organization whose enforcement power over groups that use the Porcupine caribou herd is in the form of moral persuasion (i.e., drawing on the legitimacy of traditional cultural ways of relating to caribou) and rational appeals (i.e., presenting the current state of the caribou herd in quantitative terms). The front and back sides of a PCMB pamphlet (see Figures 12.1 and 12.2) reflect this approach.

The front side demonstrates the appeal to tradition in the statement "Respecting the caribou is the best way of protecting our traditions / Leave the cows alone," and the back side details the quantitative model used to understand the state of the Porcupine caribou herd as well as the

FIGURE 12.1   PCMB pamphlet, front side | *Photo by Peter Mather*

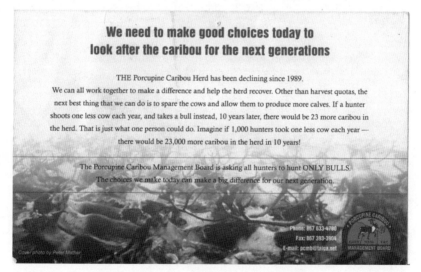

FIGURE 12.2   PCMB pamphlet, back side | *Photo by Peter Mather*

mathematical and rational explanation for why a hunter should not shoot cow caribou. In addition to presenting and describing preferred behaviours, the PCMB provides a practical rationale for suggested behaviours. For example, when discussing ways to reduce stress on caribou, they point out that snowmobiles stress caribou and result in lower-quality meat.

The PCMB's recommendations include explanations and justifications for suggested behaviour, whereas the Gwich'in rules-in-use and the territorial government's rules do not. The PCMB takes this approach likely because it has relatively little sanctioning power over its audience due to the fact that it is an advisory body. In contrast, the territorial government's regulations have the power of law behind them, and the Gwich'in rules-in-use have community-level social sanctions arising from the community's acceptance of the way things are done. The PCMB recognizes that hunting behaviours are ultimately the choice of the hunter, so it takes an educational approach. In this way, hunter autonomy is recognized. In conclusion, the three sets of rules or rules-in-use have general content in common, and the differences are found in the amount of autonomy granted by each one.

* * *

This chapter focuses on the different kinds of rules or rules-in-use that govern caribou hunting in the Gwich'in region of the Northwest Territories. Some of these are formally defined regulations set out in the Wildlife Act R.S.N.W.T. 1988, c.W-4; many more are guidelines or recommendations established by the Porcupine Caribou Management Board. Also included here are those informal rules or socio-cultural norms – rules-in-use – that are part of the traditional system of caribou management of the Teetł'it Gwich'in. These rules-in-use, although not written down in the same way as regulations and guidelines, were identified in the research through semi-structured interviews with Teetł'it Gwich'in elders.

The comparison was undertaken to better understand the degree of congruency between rules or rules-in-use at different institutional scales, from the territorial and regional levels to the community and household levels. The three sets of rules or rules-in-use are compared and contrasted, and findings include a high level of agreement between them. It is suggested that the differences have less to do with content and more to do with form and implementation. There is an increasing degree of detail and rigidity in the nature and application of rules or rules-in-use as scale increases; traditional rules-in-use defined by elders assume a high degree of flexibility and appear and function more like principles, the edges of which are defined through the exercising of hunter autonomy and responsibility (Wray & Parlee, 2013). They are flexible in providing hunters with the capacity to adapt to the characteristic variability and uncertainties of barren-ground caribou and northern livelihoods. Thus this analysis affirms the importance of understanding caribou management as a complex system of socio-cultural relations consisting of many overlapping and sometimes conflicting sets of rules or rules-in-use.

The key question for those involved in the study of co-management arrangements and for communities such as the Teetł'it Gwich'in is how the lack of fit between local rules-in-use and the rules of the territorial government and co-management boards can be resolved. Do the rules set out by co-management boards interfere with those traditionally in use by local harvesters? Or can the rules and rules-in-use of these different institutional scales co-exist? A deeper question deals with the perceived legitimacy of the Porcupine Caribou Management Board and its authority at the local level. Although much work has been done, including communication by the board about rules such as "Leave the cows alone," Gwich'in hunters were better able to identify "traditional rules" as spoken by Gwich'in elders in this study. Further, there was a good match between the kinds of rules that were identified by active harvesters and those previously identified by elders as "traditional rules for respecting caribou." Although the youngest hunters interviewed (aged 19 to 32) spoke of fewer rules and explained them slightly more simplistically, they easily identified and relayed their ideas about ways of respecting caribou in a fashion very similar to that of their elders (Wray & Parlee, 2013). Given that the PCMB, like other co-management boards, has little power of its own and thus must work to ensure legitimacy in the eyes of both the state and local communities, reconciling incongruencies between rules-in-use at the local level and rules put forward by territorial or federal government authorities will ultimately determine its fate and ability to manage the ecological and socio-economic implications of the present caribou population decline and others that may emerge in future years.

For the Teetł'it Gwich'in hunter, being a good hunter also increasingly means negotiating between these different and sometimes conflicting sets of rules and rules-in-use. One's interpretations of when, where, and in what ways these rules and rules-in-use are relevant and applicable will also affect the role and legitimacy of co-management institutions in wildlife management now and in the future.

REFERENCES

Berkes, F. (1989). Cooperation from the perspective of human ecology. In F. Berkes (Ed.), *Common Property Resources: Ecology and Community-Based Sustainable Development* (pp. 70–88). London: Belhaven.

Berkes, F., & Berkes, M.K. (2009). Ecological complexity, fuzzy logic and holism in indigenous knowledge. *Futures, 41*(1), 6–12.

Canada & Gwich'in Tribal Council. (1992). *Gwich'in Comprehensive Land Claim Agreement*. Ottawa: Indian and Northern Affairs Canada.

Davidson-Hunt, I., & Berkes, F. (2003). Learning as you journey: Anishinaabe perception of social-ecological environments and adaptive learning. *Conservation Ecology, 8*(1), 5.

Fletcher, C. (2003). Community-based participatory research relationships with Aboriginal communities in Canada: An overview of context and process. *Pimatziwin: A Journal of Aboriginal and Indigenous Community Health, 1*(1), 27–62.

Government of the Northwest Territories (GNWT) (n.d.). *Environment and Natural Resources*. http://www.enr.gov.nt.ca/en.

Government of the Northwest Territories (GNWT). (2009). *Northwest Territories: Summary of Hunting Regulations July 1, 2009 – June 30, 2010*. Yellowknife: Department of Environment and Natural Resources, Government of the Northwest Territories.

Government of the Northwest Territories (GNWT). (2011). *Caribou Forever – Our Heritage, Our Responsibility: A Barren-Ground Caribou Management Strategy for the Northwest Territories, 2011–2015*. Yellowknife: Department of Environment and Natural Resources, Government of the Northwest Territories.

Guédon, M.F. (1994). Dene ways and the ethnographer's culture. In D.E. Young & J.-G. Goulet (Eds.), *Being Changed by Cross-cultural Encounters: The Anthropology of Extraordinary Experience*. Peterborough, ON: Broadview.

Ingold, T. (2000). *The Perception of the Environment: Essays on Livelihood, Dwelling, and Skill*. London: Routledge.

Johnson, L.M., & Hunn, E.S. (2010). *Landscape Ethnoecology: Concepts of Biotic and Physical Space*. Oxford: Berghahn Books.

Kofinas, G. (1998). *The costs of power sharing: Community involvement in Canadian porcupine caribou co-management* (PhD diss.). University of British Columbia, Vancouver.

Kofinas, G. (2005). Caribou hunters and researchers at the co-management interface: Emergent dilemmas and the dynamics of legitimacy in power sharing. *Anthropologica, 47*(2), 179–96.

Kritsch, I., Andre, A., & Kreps, B. (1994). Gwichya Gwich'in oral history project. In J.-L. Pilon (Ed.), *Bridges across Time: The NOGAP Archaeology Project* (pp. 5–13). Canadian Archaeological Association.

Nadasdy, P. (2003). *Hunters and Bureaucrats: Power, Knowledge, and Aboriginal-State Relations in the Southwest Yukon*. Vancouver: UBC Press.

Nelson, R.K. (1983). *Make Prayers to the Raven: A Koyukon View of the Northern Forest*. Chicago: University of Chicago Press.

Padilla, E. (2010). *Caribou leadership: A study of traditional knowledge, animal behavior, and policy* (MSc thesis). University of Alaska Fairbanks, Fairbanks.

Pinkerton, E.W. (1989). Attaining better fisheries management through co-management: Prospects, problems, and propositions. In E. Pinkerton (Ed.), *Co-operative Management of Local Fisheries: New Directions in Improved Management and Community Development* (pp. 33–40). Vancouver: UBC Press.

Porcupine Caribou Management Board (PCMB) (2004–13). *Porcupine Caribou Management Board*. http://www.pcmb.ca.

Porcupine Caribou Management Board (PCMB). (2010). *Harvest Management Plan for the Porcupine Caribou Herd in Canada*. Yellowknife: Na-Cho Nyak Dun First Nation, Gwich'in Tribal Council, Inuvialuit Game Council, Tr'ondëk Hwëch'in First Nation, Government of Vuntut Gwitchin, Government of the Northwest Territories, Government of the Yukon Territory, and Government of Canada.

Revised Statutes of the Northwest Territories (R.S.N.W.T.). (1988). *Wildlife Act*. Yellowknife: Territorial Printer.

Sherry, E. & Vuntut Gwitchin First Nation. (1999). *The Land Still Speaks: Gwitchin Words about Life in Dempster Country*. Old Crow, YT: Vuntut Gwitchin First Nation.

Wray, K., & Parlee, B. (2013). Ways we respect caribou: Teetł'it Gwich'in rules. *Arctic, 66*(1), 68–78.

13

Letting the Leaders Pass: Barriers to Using Traditional Ecological Knowledge in Co-management as the Basis of Formal Hunting Regulations

Elisabeth Padilla and Gary P. Kofinas

My mama always tell us don't bother 'em. Let them go through first, there be more caribou coming. And if they have a trail, caribou follow. She said don't bother 'em. So we always respected, we never bother the front caribou.

– Old Crow hunter

In this chapter, we explore a case where an attempt to apply traditional ecological knowledge (TEK) in co-management as the basis for formal hunting regulations resulted in failure. The case in question – the "let the leaders pass" policy, which was established for the Dempster Highway of the western Canadian Arctic and applied to the Porcupine caribou herd – allowed us to identify the conditions that create barriers to the successful application of TEK through co-management. We analyze the case through four historical phases of caribou management and complement the study with a literature review of TEK and wildlife co-management to explain why TEK integration of caribou leaders in regulatory resource management fell short of success. Identifying and understanding the social dynamics related to these barriers suggest solutions for transforming the co-management process.

We use the term "traditional ecological knowledge" instead of "Indigenous knowledge" or "local knowledge," despite ambiguous definitional issues, because it is more specific to Indigenous people's knowledge of the land (Berkes, 1993, 2012). We regard "TEK" as "a cumulative body of knowledge, practice, and belief, evolving by adaptive processes and handed down through generations by cultural transmission, about the relationship of living beings (including humans) with one another and with their environment" (Berkes, 2012, p. 7). We define "co-management" in multiple ways, as reviewed by Carlsson and Berkes (2005). Their review emphasizes that co-management has complexities, variations, and a dynamic nature that are not captured by most definitions. We chose to use Singleton's (1998, p. 7) definition because it best reflects the ideals of co-management in this case study: "governance systems that combine State control with local, decentralized decision making and accountability and which, ideally, combine the strengths and mitigate the weaknesses of each."

CARIBOU LEADERS AND "LETTING THE LEADERS PASS"

The notion of letting the caribou leaders pass originates in Indigenous hunting tradition. TEK describing caribou leaders (i.e., caribou that lead a group or herd of caribou) is sparsely documented for Indigenous caribou-hunting groups. Gubser (1961) describes caribou leaders in his early ethnography of Nunamiut Eskimo at Anaktuvuk Pass, Alaska. Porcupine caribou hunters traditionally avoid shooting caribou leaders to ensure continued migration of caribou and caribou harvest throughout the fall hunting season (Sherry & Vuntut Gwitchin First Nation, 1999; Smith & Cooley, 2003). Traditional hunting strategies of Inuit in the Northwest Territories show concern for deviating caribou movement from common migration routes. For example, Gunn, Arlooktoo, and Kaomayok (1988) describe hunters remaining inconspicuous at water crossings until lead caribou cross so that following caribou will continue to swim across when hunters begin to shoot. Similarly, Stewart, Keith, and Scottie (2004) report that hunters avoid butchering animals or disposing of bones near migration routes to avoid diverting caribou movement.

The tendency of a group to follow the first "leader" animal that leaves the herd is well documented for reindeer (Baskin, 1989; Paine, 1988). Ethnographic accounts of reindeer herding, such as those of the Saami of Fennoscandia, highlight herders' knowledge of herd leaders and their importance in directing the movements of groups of *Rangifer* (e.g., Paine, 1994). The identification and use of leader animals to facilitate herd management is central to the practice of reindeer herding (Baskin, 1989).

Baskin and Hjalten (2001) describe the tendency of a group of reindeer to run away following a leader and herders consequently eliminating the most nervous animals to ease herd management. Additionally, herders will use docile animals as decoys to lead a group of domestic reindeer to a new pasture or to lead a group of wild reindeer into an ambush (Ingold, 1986).

Few scientific studies have measured or systematically observed caribou leadership behaviour (Benn, 2001; Miller, Jonkel, & Tessier, 1972; Paine, 1988). There is, however, a growing body of literature on leadership behaviour in various species, with potential applications to population management (Bailey, Dumont, & WallisDeVries, 1998; Conradt & Roper, 2005; Couzin, Krause, Franks, & Levin, 2005; Dumont, Boissy, Achard, Sibbald, & Erhard, 2005; McComb, Moss, Durant, Baker, & Sayialel, 2001; Rands, Cowlishaw, Pettifor, Rowcliffe, & Johnstone, 2003). Some have argued that the expanding human infrastructure has had and/or will potentially have a negative effect on caribou (Nellemann, Vistnes, Jordhøy, & Strand, 2001). Although scientific investigation of caribou leadership is important to understanding the impact of human infrastructure and related activities on caribou movement (Benn, 2001; Smith & Cooley, 2003), leadership in caribou has not been measured using the scientific method. Authors have, however, described these gregarious animals as having well-established, learned migratory behaviour characterized by fidelity to calving grounds or crossing points (Dahle, Reimers, & Colman, 2008; Kelsall, 1968). Caribou leaders could thus exist for migration purposes and/or within temporary groups. When the Porcupine Caribou Management Board deliberated on how to address concerns about the effects of the Dempster Highway on the harvesting of caribou leaders, the body of scientific knowledge was limited.

Indigenous residents of rural communities in the Yukon and the Northwest Territories have expressed concern that disturbing or shooting caribou leaders can cause caribou to redirect seasonal movements and abandon wintering areas in proximity to communities (Benn, 2001). Similarly, Smith and Cooley (2003) report hunters' predictions of caribou reactions to roadside disturbances. Hunters in that study based their answers on their own interactions with caribou. They found that "taking the leaders" was a predictor of how caribou react to human disruption. Participants in the study mentioned caribou leaders when talking about the location of caribou killed in a group. Respondents in that study described types of caribou leaders, including individuals or groups of animals with different behavioural roles in the herd, such as steering migrating groups or detecting danger.

The practice of letting the leaders pass is still supported by many subsistence communities that rely on northern caribou and by some Indigenous hunters, and elders have brought the concept of caribou leaders to the forefront of management discussions (e.g., Alaska Department of Fish and Game, 2005; PCMB, 1995). "Caribou leaders," however, are not easily defined as one particular sex or age class. Instead, "leaders" is a flexible term used in ways directly applicable to hunting traditions, and it is therefore highly dependent on social-ecological context (Padilla, 2010). As discovered in this study, creating a formal hunting regulation based on this traditional practice is a challenge.

Managing Resources with TEK

Many authors have argued that incorporating TEK into resource managers' decision making can increase resource management effectiveness (e.g., Berkes & Turner, 2006; Chapin, Kofinas, & Folke, 2009; Folke, 2004; Houck, 2003; Stevenson, 1996). Early frames of "systems of knowledge" were typically dichotomized (i.e., Western vs. traditional knowledge), with the opportunities for integration simplistically described. Several co-management bodies of Canada were established in the late 1970s and onward to involve Indigenous resource users in the wildlife management process, with some arrangements explicitly recognizing the legitimacy of local and traditional knowledge (Houde, 2007). Subsequent analyses have offered more nuanced descriptions of the complexities of the process and its political underpinnings (e.g., Berkes, 2012; Huntington, 2005; Kofinas, 2005; Morrow & Hensel, 1992; Parlee, Manseau, and Łutsël K'é Dene First Nation, 2005). Reframing the discussion on TEK to focus on social learning, Berkes (2012) describes traditional knowledge and resource management as processes of "trial and error" and "learning by doing" while managing for the unknown. The recognition of multiple sources of knowledge in management indicates the potential for resource users, managers, and scientists to work together in an adaptive co-management system for a more holistic understanding and for more robust policies (Armitage, Berkes, & Doubleday, 2007; Kendrick, 2003). Moving from theoretical speculation to application raises the question of whether, when, and how traditional knowledge can contribute to wildlife management. This chapter helps to address that question through a study of the use of TEK in formal wildlife regulations and its implementation through a co-management process.

In 1986 the Government of Canada established the Porcupine Caribou Management Board (PCMB) through the Porcupine Caribou Management Agreement (Government of Canada et al., 1985), which was signed by the Government of the Yukon Territory, the Government of the Northwest Territories, and First Nations of the region (Kofinas, 1998; Peter & Urquhart, 1991). The terms of the agreement state that the board is an advisory body charged with making recommendations to government ministers. Although the authority of the board is therefore limited, elected officials and agencies have deferred to the board's recommendations because of its broad representation and its role as a communication and coordination body (Kofinas, 1998). The creation of the agreement and its board was an effort to link local-level authority systems with higher-level processes of decision making and to more effectively achieve regional consensus on caribou management decisions (Kofinas, 2005).

One key motivation in establishing the Canadian arrangement for co-managing the Porcupine caribou herd was to address longstanding issues related to hunting Porcupine caribou along the Dempster Highway, a 736-kilometre, two-lane gravel road running from the central Yukon to Inuvik in the Northwest Territories. Upon the opening of the Dempster Highway in 1979 (Page, 1986), a range of wildlife management issues was of concern to both government officials and hunters, including safety over hunting from the Dempster Highway because of easy access, the potential abandonment of traditional hunting practices through highway hunting with trucks, and disturbance to the herd. Among the concerns was the worry that highway hunting was causing a deflection of the caribou herd's migration from wintering grounds east of the highway, thus making caribou unavailable to local hunters of the Northwest Territories and the Yukon. Based on a recommendation to the PCMB from the local Tetlit Gwich'in Renewable Resource Council of Fort McPherson in 1994, TEK was later used by the Governments of the Yukon and the Northwest Territories as the basis for hunting closures along the highway to let the leaders of the caribou herd pass (PCMB, 2000). The absence of science-based recommendations on caribou leaders and the complete reliance on TEK as the basis for the hunting regulation set this hunting closure apart from the other hunting regulations.

Exploring Co-management Dynamics of "Letting the Leaders Pass"

Achieving effective co-management is a collective-action problem among stakeholders, including government agencies. Understanding the dynamics of the co-management process for the "let the leaders pass" regulation suggests the need to consider the costs and benefits of various actors at various scales, from the individual hunter to communities, First Nations, and territorial and federal governments, as well as the historical context from which the regulation emerged. As noted by several previous studies, co-management is typically no panacea (Caulfield, 1997; Caulfield et al., 2004; Kofinas, 1998, 2005; Nadasdy, 2003b; Natcher, Davis, & Hickey, 2005). Despite early theoretical hopes of northern wildlife co-management leading to more effective resource management (e.g., Osherenko, 1988; Usher, 1986), the political conflicts of co-management have in some cases led to significant problems (e.g., Nadasdy, 2003b). Suggested in all these case studies is that the actions and reactions of governments, scientists, local community leaders, and hunters at different scales are a key consideration in the analysis of emergent co-management challenges.

The "let the leaders pass" policy implementation differed between the Yukon Territory and the Northwest Territories, where it was not formally enforced. Our focus in this chapter is on the Yukon Territory's implementation of the hunting regulation established in 2000 and its eventual rejection by the federal Department of Justice after it was challenged by a Dawson City First Nation hunter in 2007. Table 13.1 shows how, since its first meeting in 1986, the PCMB has engaged in an ongoing process of board-level discussions, community consultations, media coverage, and workshops to address and reassess problems associated with Dempster Highway hunting and how the PCMB ultimately proposed "let the leaders pass" hunting regulations.

We explored the case through a historical analysis of Dempster Highway hunting issues, dividing the events into four time periods or phases:

- Phase one: traditional management (precontact to 1950s)
- Phase two: pre–Porcupine caribou herd co-management, Dempster Highway construction to completion, and signing of the Porcupine Caribou Management Agreement (1960s–85)
- Phase three: early Porcupine caribou herd co-management through the Porcupine Caribou Management Board (1985–95)
- Phase four: later Porcupine caribou herd co-management with traditional knowledge in a new political context (1995–2009)

TABLE 13.1 Phases of caribou management with respect to the Dempster Highway and changes in social organization, role of traditional ecological knowledge (TEK), and wildlife management relations

Phase	Social organization of caribou harvesting by Fort McPherson residents	Role of TEK in wildlife management	Wildlife management
One: traditional management (precontact to 1950s)	Community hunts common, with chief or hunting leader directing actions of hunters. Kinship groups and individuals also harvesting opportunistically. Use of foot and snowshoes for mobility.	Cultivated and highly applied by Indigenous hunters. Some limited interest by naturalists and anthropologists.	Traditional system of caribou management, with state-community interactions in caribou management limited.
Two: pre–Porcupine caribou herd co-management (1960s–85), Dempster Highway construction to completion, and signing of the Porcupine Caribou Management Agreement (1960s–85)	Dog sleds replaced with snowmobiles by the 1960s, with hunters using trucks to access hunting areas around the Dempster Highway. Greater individualism in harvesting.	TEK evolves at local level, whereas state management mostly dismisses the value of TEK and seeks scientific insights for caribou management.	Increased presence of state-dominated caribou management. State agencies "consult" with communities. Little to no power sharing in decision making.
Three: early Porcupine caribou herd co-management through the Porcupine Caribou Management Board (1985–95)	Dempster Highway corridor becomes most common access point for caribou hunting. Greater schism between youth and elder perspectives on ethical hunting.	TEK gains legitimacy as co-management develops. First TEK studies conducted during this time.	Porcupine caribou herd co-management system established. State agencies increasingly look to PCMB for solutions to issues and are responsive to board recommendations.
Four: later Porcupine caribou herd co-management with traditional knowledge in a new political context (1995–2009)	"Let the leaders pass" articulated by leaders, recommended to the PCMB, and instituted in the Yukon.	TEK embraced by board and used as basis for addressing Dempster Highway disturbance concerns but later challenged by Indigenous hunters.	State management is highly responsive to PCMB recommendations.

We describe the social-political context of Indigenous caribou-hunting management from the early system of traditional authority to the recent PCMB co-management arrangement, and we identify barriers that prevented successful application of TEK within regulatory wildlife management. Stated as middle-range propositions, these barriers include:

- The context-specific nature of TEK limits its application in resource management regulations.
- Changes in systems of traditional authority, hunting technology, and the social organization of harvesting caribou affect the effectiveness of TEK approaches in a contemporary social setting.
- Indigenous efforts toward self-government and political autonomy limit regional co-management consensus in a heterogeneous cultural landscape.
- The mismatch of agency enforcement of hunting regulations and TEK-based education is problematic.

Our analysis is complemented with a literature review of TEK and wildlife co-management to explain why TEK integration of caribou leaders in regulatory resource management fell short of success.

METHODS

Case Study Research

We used the approach of a single case study to analyze the "let the leaders pass" policy. As defined by Yin (2009, p. 18), a case study is "an empirical inquiry that investigates a contemporary phenomenon in depth and within its real-life context, especially when the boundaries between phenomenon and context are not clearly evident," and a case study can provide insight into specific complex phenomena where experiential research and controlled variables are not possible. Several researchers in the past have used single case studies to assess the effectiveness of integrating TEK into government-led resource management (e.g., Berkes, 2012; Kendrick, 2003; Nadasdy, 2003b; Richard & Pike, 1993; Taiepa et al., 1997). Pinkerton (1989) elucidates the conditions for successful co-management in her seminal analysis and other studies (Pinkerton, 1994, 2009). She argues that identifying the nature of barriers and ways to overcome them in smaller-scale case studies of co-management implementation is a valuable method of analysis that can contribute to a broader understanding of co-management systems (Pinkerton, 1999). Similar to Light, Gunderson, and Holling's (1995) historical account of the Everglades, we use a "thick description" (Geertz, 1973) to present data or key events relevant to the case study. We follow a grounded-theory approach to studying co-management (Glaser & Strauss, 1967; Strauss, 1987) in order to identify barriers to implementing TEK-based hunting regulations through co-management. Because of the use of a single case study, we note from the outset that there are inherent limitations in drawing broader generalizations about the phenomena. This suggests a need to test our four propositions in other wildlife co-management contexts.

Sources of Evidence and Historical Account

We studied the case of Indigenous caribou hunting from the Dempster Highway using multiple sources of evidence and, where possible, sought to triangulate evidence in order to confirm factual information (see Table 13.2). Research for the case study was conducted through two projects, with findings from both integrated into this analysis. The first was a project funded by the Man and the Biosphere Reserve program of the National Science Foundation (Kofinas, 1998), which examined the cost of power sharing in co-management. The second was the master's thesis research of Padilla (2010), which focuses on local hunters' definitions of caribou leaders and their perspectives on Dempster Highway hunting regulations. The former offered a foundational understanding of Porcupine caribou herd co-management and insight into historical events surrounding the "let the leaders pass" policy. The Kofinas study involved extensive archival research, three years of participant observation in caribou-user communities, and participant observation at ten PCMB meetings from 1992 to 1997, including one in 1995 when the "let the leaders pass" idea was first presented to the PCMB as a proposal by an Indigenous leader. This project included unstructured interviews with key respondents, including all PCMB members and many community leaders, and over 220 structured interviews with Porcupine caribou herd users from communities, as well as focus group research with Fort McPherson hunters about the Dempster Highway hunting. In addition, Kofinas attended several PCMB-organized workshops focused on the Dempster Highway hunting problem.

Padilla's research on caribou leaders and highway hunting provided a more contemporary account of events and perspectives during the final

TABLE 13.2   Sources of data

Source	Period	Details
Participant observation in co-management process, including observations of PCMB meetings	1992–98, 2005, 2006	Kofinas lived in user communities for ten months, attended meetings, travelled with PCMB members who were hunting on the Dempster Highway with local residents, and participated in all workshops during the 1992–98 period. Padilla attended three PCMB meetings, in addition to being in communities during interviews.
Focus group	March 1995	15 Fort McPherson hunters participated.
Archival research on Dempster Highway and Dempster Highway hunting	1995–2004	Yukon and Northwest Territories agency personnel provided access to agency files, and other searches were conducted at the Yukon Archives of Yukon College.
Interviews of caribou users on co-management	1994–96	220 structured interviews completed and analyzed.
Coding of PCMB minutes	1986 (first PCMB meeting) to 2007	Coded thematically to capture issues and actions.
Semi-structured interviews with PCMB members, community leaders, and agency personnel	1994–97, 2005	20 interviews completed, transcribed, and coded.
Interviews with elders and caribou hunters on caribou leaders and the "let the leaders pass" rule	2006	29 interviews completed with elders and hunters of Old Crow (9), Dawson (9), and Fort McPherson (11).

period, when the "let the leaders pass" policy was enforced and then rejected. Interviews on caribou leaders and hunting were conducted in three First Nation communities in the Yukon and the Northwest Territories that are among the primary users of Porcupine caribou. They included Dawson City (population 1,327), Fort McPherson (population 776), and

Old Crow (population 253) (Statistics Canada, 2006). This research was conducted in partnership with the Tr'ondëk Hwëch'in First Nation, based in Dawson City, the Tetlit Gwich'in Renewable Resource Council of the Gwich'in First Nation, based in Fort McPherson, and the Vuntut Gwitchin First Nation, based in Old Crow. These three communities cover a wide geographic area along the Porcupine caribou's migration and wintering range, and research in the communities showed their involvement in issues related to Dempster Highway hunting, as well as their differing social-ecological characteristics, such as population or road access. Fort McPherson and Dawson hunters make extensive use of the Dempster Highway to access Porcupine caribou, whereas Old Crow hunters primarily use river and snowmobile access, depending on the season. Fort McPherson hunters take the greatest number of caribou of all the communities of Canada and Alaska that hunt Porcupine caribou, and Old Crow harvests the highest number of caribou per capita of all the communities (Kofinas, 1998). Although Dawson harvests caribou, the community is more dependent on moose. Of the 29 interviews with hunters of the Porcupine caribou herd and with elders during the summer of 2006, 9 were completed in Dawson City, 11 in Fort McPherson, and 9 in Old Crow. In each community, we hired and worked with a local assistant, who worked with the local First Nation or renewable resource council to select interviewees. All respondents in these communities were members of the local First Nation. Both younger and older hunters were selected not only to capture changing knowledge (Stevenson, 1996) but also to capture a wider space-time scale (Ferguson & Messier, 1997). Local organizations advertised this study of caribou leaders for several weeks before our arrival to inform the community about the project. Local assistants contacted specific respondents informally to ask for their participation in interviews shortly before they were conducted. We interviewed 5 women and 24 men. Their ages ranged from 42 to 91, with 16 being elders. We called the rest of the respondents "younger hunters."

Interviews on caribou leaders and hunters were semi-structured, and most questions were open-ended to allow subjects the freedom to respond with detailed information and anecdotal personal accounts (Huntington, 2000; Perecman & Curran, 2006). Interviews were constructed around the assumption that traditional knowledge is rooted in local cultural contexts and arises from long-term interactions between people and resources (Berkes, 2012; Folke, 2004). We audio-recorded all interviews and transcribed their scripts entirely. We used HyperRESEARCHTM

coding software for qualitative analysis and a content-analysis protocol through open coding. Open coding provided the basis for arriving
at categories of responses that were captured as quotations within each
interview.

Together, the research findings of the Kofinas and Padilla projects provided a uniquely rich source of evidence for understanding the historical
and multiscale transactions associated with Dempster Highway caribou
hunting and the nature of the failed "let the leaders pass" policy.

Historical Analysis of Phases of Management and Identification of Barriers

Several research questions guided our identification and review of the four
phases of caribou management. They include who managed Indigenous
caribou hunting; what were the management strategies; and how, when,
and for how long were they implemented? Through extensive archival
research and interviews with key informants, we reconstructed historic
events by reviewing printed media communications, accessing internal
agency and board memos, and coding co-management board minutes and
other documentation. We describe the management strategy under which
the "let the leaders pass" policy was implemented and why it was no longer
enforced. Throughout the account, we describe the relevant interactions
between different levels of government, communities, and hunters. This
factual account serves to illustrate the context that led to the struggle of
interests between the co-management board, government agencies, communities, resource users, and stakeholders' cultural values and management
approaches.

We identified barriers by classifying conditions from the historical
account under general themes – such as worldview aspects of traditional
knowledge, problems of scale, institutional dynamics, social characteristics, and so on – based on a review of a wide range of literature,
including writings on traditional knowledge and co-management. We
identified comments and themes from the formal interviews to illustrate
the discussion. We then grouped these themes and conditions to generate
aforementioned "barriers" specific to the case. We later shared these barriers with key respondents and modified them based on their review. A
detailed presentation of the interview analysis to identify caribou leaders
was undertaken by Padilla (2010).

Historical Phases of Dempster Highway
Hunting Issues and Management

Phase One: Traditional Management (Precontact to 1950s)

Slobodin (1962), McClellan (1987), Frank and Frank (1995), Kofinas (1998), and others have described the local system of caribou management for communities that hunt the Porcupine caribou herd, including informal institutional arrangements shaping community decision making related to hunting. From precontact with westerners to the mid-1900s, hunting commonly occurred as community hunts under the direction of a chief or leader (Slobodin, 1962). Before the adoption of the rifle, people used "caribou surrounds," or fences, to direct herds into harvest areas, with family groups working cooperatively to achieve a successful hunt (Greer & LeBlanc, 1992; Warbelow, Roseneau, & Stern, 1975). Hunting caribou successfully was a matter of survival during precontact, demanding that the community members work cooperatively to intercept, kill, butcher, and distribute caribou meat. Old stories, as reported by Gwich'in elders, tell how caribou and people were one and the same during "the time before there was time" and how, after the separation of caribou from people, caribou gave themselves to hunters as gifts, with each partner fulfilling a set of obligations to care for the other (Kofinas, 1998; Slobodin, 1981). From these understandings of human-environment relations derived norms for respectful behaviour toward animals, such as proper methods of hunting, obligations to share the harvest with others, and sanctions against disrespect or wastage of meat (Sherry & Vuntut Gwitchin First Nation, 1999). Many of those traditional principles of conduct for hunting are present today in the worldview and behaviour of Indigenous users of the Porcupine caribou. Elders interviewed in this study spoke of the traditional harvesting method as a model for good hunting and compared it with the less structured modern hunting:

> Around '50s before that, you, well people used to move, you see, and I seen it, too, because when you move, there was one guy, chief, always one guy that talked in the morning. You don't just go out there, you can't. So the guy that's in charge, and this is really morning, too, as soon as you get up after dark, you got to get up. And he tell you got to go this way today, you go this way, and you go there. Sometime when there's lots of people, another bunch go there. If you see caribou, you don't just shoot it, have to come back, caribou going to come back. The herd. So the next day, you start planning that night. Next day, you go there and everybody hunt.

That's way you do it, you don't just go hunting ... Should be certain time you go up on the highway. There's people with every truck. Should be one boss, and then there wouldn't be a problem. (Fort McPherson respondent)

We had one boss that we go out hunting. We all gather together, and the oldest guys told us what to do. You go there, you go there, you go there, so everybody knows who's here, who's there. That's the way we used to do it ... We used to have circle around the caribou. Now you just go up and help yourself and don't have to wait for nobody. (Old Crow respondent)

Beginning in the late 1950s through the 1970s, there was a significant change in the social organization of traditional caribou hunting because of the introduction of new harvesting technologies; hunting by snowmobile and automobile mostly replaced hunting by foot, snowshoe, and dog sled. A hunter described the change from traditional hunting and spending considerable time on the land to the practice of using trucks for shorter hunts:

People never used to hunt in town like this, like nowadays. They were gone for couple weeks sometime, a month, they just live out there. Nowadays they hunt with trucks, they don't even skin caribou up, they bring them back like that. Nowadays the generation is different from long ago, way different, way, way different. More modern than long ago. (Fort McPherson respondent)

The traditional organization of hunting caribou described by these elders illustrates how leadership and social controls on hunting behaviour were linked and coordinated. Caribou hunting later became a more individualized activity. Although small group hunts are still common, the community hunt under the leadership of a skilled "boss" became an anomalous means of harvesting, and the role of the hunting leader in overseeing group hunting mostly disappeared.

Phase Two: Pre-Porcupine Caribou Herd Co-management, Dempster Highway Construction to Completion, and Signing of the Porcupine Caribou Management Agreement (1960s–85)

In 1959, Canada's prime minister, John Diefenbaker, announced the construction of the "the road to resources," which would connect the southern Yukon Territory with anticipated hydrocarbon development activities in the Mackenzie Delta region of the Northwest Territories. Completed in 1979, the portion of the Dempster Highway at the border of the Yukon and the Northwest Territories linked the Yukon community of Dawson City to the Northwest Territories communities of Fort McPherson, Tsiigehtchik (formerly Arctic Red River), and Inuvik.

The indirect effects of the highway and hunting on the highway posed the greatest concern. Early scientific studies investigated the impact of the highway and traffic on caribou migration but found little to no evidence of disturbance (Dempster Highway Working Group, 1978). In contrast, a panel of biologists convened by the federal government predicted that increased access and hunting from the Dempster Highway would reduce the size of the Porcupine caribou herd by 30 to 40 percent (Alaska Highway Pipeline Panel, 1979). Articles from the local media in the mid- to late 1970s also warned of the impact of Dempster Highway hunting on caribou, with headlines such as "Highway could deplete herd" (Fraser, 1976) or suggestions such as "Improper management of the area could result in the herd's extinction, according to many experts" (Renaud, 1978). The concerns regarding the future of the herd were also elevated to a national and international level through the highly publicized Mackenzie Valley environmental impact assessment chaired by Justice Thomas Berger, known as the Berger Inquiry. The impact assessment and its community hearings captured the attention of many Canadians and brought a heightened awareness of the Porcupine caribou herd and the potential impacts of Dempster Highway hunting (Berger, 1977).

A lack of consultation with Indigenous hunters characterized the period prior to co-management. For example, the chief of Fort McPherson during the final construction phase of the highway reported to Kofinas that his community learned about the plans for highway construction only indirectly. When the community's leadership was told by government that the highway would be constructed, the government stated that it would be "giving the people a road." He and other elders in the community reported that most people in the community had little idea of the implications of a road and that community leadership had no say in the highway's construction or routing.

Until 1985 the Government of the Yukon Territory was responsible for regulating hunting and based wildlife management decisions primarily on scientific studies (McCandless, 1985). The paradigm of state-controlled wildlife management and the dismissal of Indigenous knowledge were evident in internal Yukon government memoranda between the director and assistant director of wildlife in the late 1970s. In 1978 the Yukon government's Department of Renewable Resources formed a Dempster Highway committee to assist in coordination of the Porcupine caribou herd's management. Its jurisdiction extended only to non-Indigenous hunters, unless the species was declared endangered. A particular focus of the committee was finding ways to limit Indigenous hunting because these users constituted the majority of caribou harvesters.

The absence of recognized rights by Indigenous hunters of the Porcupine caribou herd was discussed by top Yukon government officials, with the Yukon government's assistant director of wildlife questioning whether Indigenous people should be given special rights to hunt caribou, as he did not recognize the need to hunt caribou as critical to survival in contemporary times:

> Perhaps the question of Native hunting rights could be evaluated. In my opinion, there is no dependence anymore on caribou to survive. If caribou are not shot because they don't come close to a community, meat is being made available through other means. In light of this, I question the philosophy that this herd should be managed primarily for Native use and to perpetuate traditional lifestyle.

Indigenous hunters shifted their perspective about whether formal regulations were applicable to them. After an abnormal calving year reported in 1977, continued predictions of a declining population sensitive to subtle increases in harvest, and observations of changes in the migration route, biologists urged a closure of the road to hunting. Their recommendations to government led to the establishment of a sixteen-kilometre no-hunting corridor along the highway in 1978. The restrictive corridor had the desired impact, as hunting dramatically decreased in subsequent years, partly through the compliance of Indigenous hunters who perceived that the regulation superseded constitutionally based Aboriginal hunting rights. However, from 1981, the general perception changed, with the corridor hunting restrictions being viewed as legally not applicable to Indigenous people, and caribou hunting from the Dempster Highway again increased in the Fort McPherson area. The hunting corridor was then decreased to two kilometres, causing Indigenous harvest along the southern portion of the Dempster Highway to increase. These multiple changes to the policy on a no-hunting corridor were the first of continuous changes in hunting management from then onward.

Preferential consideration for opinions voiced by residents of Old Crow, an off-road community that is highly dependent on the Porcupine caribou herd, may have set the stage for increased intercommunity conflict through new efforts at range-wide caribou management. Residents of Old Crow echoed biologists' concerns over hunting caribou on the highway. Old Crow's location in the Yukon Territory gave it a political advantage over non-Yukon communities that hunted the Porcupine caribou herd, such as Fort McPherson. Government agencies perceived communities in the Yukon Territory as having little to no impact on the caribou population because of their smaller human population and more traditional style of harvesting, namely not road hunting, as evidenced in internal agency memoranda. The commissioner to the Yukon Territory, head of the government at that time, along with other government officials, sided with Old Crow community concerns in a memorandum to the chairman of the Council for Yukon Indians: "We believe that the general consensus in Old Crow is that all hunting [from the Dempster Highway] be stopped and it is our conviction that these people [Vuntut Gwitchin of Old Crow] are most closely tied to the Porcupine Caribou population." In contrast, increasingly publicized issues of unethical hunting and stories in the media of caribou slaughter by Fort McPherson hunters tainted the image of Northwest Territories hunters throughout the 1980s. Decades later, some residents of Yukon communities still hold this perception.

The Berger Inquiry and others suggested that regional solutions to management of the Porcupine caribou herd be considered. In 1983 the Wildlife Advisory Committee of that region proposed a range-wide management plan to increase cooperation with Indigenous hunters and establish harvest quotas. After years of negotiation, the Governments of the Yukon and the Northwest Territories and First Nations of the region (Kofinas, 1998; Peter & Urquhart, 1991) signed the formal Porcupine Caribou Management Agreement in 1985. The agreement, like other formal co-management agreements of its time, was signed to provide shared authority between Indigenous user groups and government agencies in decision making (Armitage, Berkes, & Doubleday, 2007; Houde, 2007; Kofinas, 2009; Pinkerton, 1989). As a part of the agreement, the Porcupine Caribou Management Board was established.

Phase Three: Early Porcupine Caribou Herd Co-management through the Porcupine Caribou Management Board (1985–95)

One of the PCMB's primary goals, as stated in the agreement, is to "cooperatively manage to ensure conservation of the PCH [Porcupine caribou herd] with the view to ensure continued subsistence usage" (Government of Canada et al., 1985). According to government negotiators of the agreement, which was created during a period of wildlife management devolution across the North (Usher, 1986), the PCMB was established in large part to address the issue of Dempster Highway hunting management (Kofinas, 1998; Kofinas, Osherenko, Klein, & Forbes, 2000; Government of Canada et al., 1985). With its representatives of First Nations of the Yukon Territory and the Northwest Territories, the Inuvialuit, non-Indigenous users, two territorial governments, and the federal government, the PCMB grappled with problems associated with the Dempster Highway from the time of its inception. Since the Canadian Porcupine caribou co-management arrangement was established, conflicts between government agencies and communities have decreased. However, Yukon communities, perceived as "more traditional" and having a low impact on the Porcupine caribou herd, and Northwest Territories communities on the highway, perceived as hunting unethically, became increasingly polarized. Intercommunity conflicts about Dempster Highway hunting increased as groups sought to jointly manage caribou hunting.

Cumulative impacts on caribou migration of human activities along roads continued to be of concern to scientists (Wolfe, Griffith, & Gray Wolfe, 2000). However, the Porcupine caribou herd had not suffered the drastic decline predicted in 1979, despite highway hunting. Instead, the herd grew until its population peaked in 1989 at 178,000 caribou (PCMB, 2010). Population-level effects related to hunting were supplanted by the greater concern over displacement of caribou from winter grounds due to highway hunting disturbance (PCMB, 1995, 2000, 2006, 2007; Smith & Cooley, 2003).

Hunter education efforts at the local level brought only limited success. An original member of the PCMB, who was also an elder and the traditional chief of Fort McPherson, took a leading role in addressing unethical Dempster Highway hunting at the local level by patrolling the highway, talking with hunters, and cleaning unsightly kill sites. The effectiveness of his efforts in the new context of post-traditional, individualistic highway hunting was, however, self-reported as limited (phase three). In addition to these initiatives, the PCMB also produced a video on the problems of Dempster Highway hunting and distributed it to residents of local user communities in an effort to increase public awareness and find a solution.

Since the PCMB's creation, recommendations for addressing Dempster Highway issues have sought to strike a balance between hunter compliance, representation of traditional knowledge, and protection of the caribou herd, producing several modifications in formal policy. Two major sets of recommendations regarding Indigenous hunters were implemented by the Yukon government in 1990. Snowmobile access was permitted for Indigenous hunters beginning on November 1, and a two-kilometre no-hunting corridor was enforced. In spite of ongoing efforts to address Dempster Highway issues, conflict over unethical hunting practices and lack of consensus among Indigenous hunters persisted.

Phase Four: Later Porcupine Caribou Herd Co-management with Traditional Knowledge in a New Political Context (1995–2009)

The "let the leaders pass" policy was proposed in 1995, at a time when the idea of integrating TEK with co-management was gaining international popularity as a solution to complex resource management problems (Berkes, 1993; Inglis, 1993; Stevenson, 1996). After local deliberations by community elders, hunters, and leaders, the Tetlit Gwich'in Renewable Resource Council in Fort McPherson sent a representative to the spring meeting of the PCMB, requesting that it implement a closure of highway hunting upon the first arrival of caribou each fall. The local council representative explained that the basis of the proposed closure was his elders' traditional knowledge about not disturbing caribou leaders during herd migration. This reference was the first time that some non-Indigenous members of the PCMB had heard of the traditional rule, which generated backstage discussions among those board members about the significance of making policy based on TEK not backed by science.

As a follow up, later in 1995 the PCMB organized a Dempster Highway Workshop with Fort McPherson hunters and elders and government-employed resource managers. The workshop report noted that because science-based wildlife studies on caribou leaders and Dempster Highway impacts on herd migrations were both costly and not available, the TEK of Indigenous elders should therefore be used as the basis of decision making (PCMB, 1995). At a subsequent meeting, the PCMB endorsed the Tetlit Gwich'in Renewable Resource Council's alternative management approach and recommended the "let the leaders pass" closure to government ministers. The Yukon Territory government, in turn, put the closure into law. The one-week hunting closure was initially applied to the entire length of the Dempster Highway and was later separated into two different closure periods to correspond with different north and south migration timing (PCMB, 2006). The PCMB and the partner management agencies had hoped the "let the leaders pass" regulation would be embraced by local hunters because it was conveyed to the board as a traditional practice. The formal regulation was formulated as a one-week closure, with dates determined to match the time when caribou first migrated near the highway, thus ensuring the caribou leaders would have time to pass.

However, local perspectives on the number, sex, and age of those caribou leaders that should be allowed to pass were not well defined (Smith & Cooley, 2003; Padilla, 2010). In interviews, local hunters expressed differing opinions about the idea of a closure, with some recommending a longer closure of up to a month, whereas others preferred to use individual judgment about whether leaders had already passed, meaning that the highway should not be subjected to a closure (Padilla, 2010). In addition, in a 2006 review of Dempster Highway regulations initiated by the PCMB, the Yukon minister of environment received mixed support from affected Yukon Territory and Inuvialuit communities for a proposed second hunting closure (PCMB, 2007).

During a meeting of the PCMB from September 22 to 24, 2007, the PCMB chair noted that honouring elders' concern for protecting caribou leaders was a priority. The PCMB emphasized the notion of respectful hunting in its media and education outreach. Interviews with elders and hunters suggested intergenerational differences in perceptions, with elders expressing dissatisfaction over modern hunting practices, whereas younger hunters reported that they hunted according to their elders' teachings.

Despite general support for the regulation, the actions of one First Nation resident from one community challenged and ultimately discontinued the regionally implemented regulation. In September 2007 a "stay of proceedings" against a Tr'ondëk Hwëch'in (Dawson) First Nation member ended enforcement of the one-week closure to let the leaders pass. This younger hunter who violated the hunting closure was charged by wildlife officers and announced his intent to contest his case in court. During the interview we conducted with him, he explained his dissatisfaction with the regulation as an infringement on his Indigenous hunting rights, asserting that consultation done to formulate the new regulation did not adequately recognize traditional knowledge of his Dawson First Nation members. Out of nine elders and hunters we interviewed in Dawson, three hunters said that the hunting closure did not match their elders' teachings, and another four elders and hunters were in partial disagreement with letting the leaders pass. Based on these interviews and informal discussions with PCMB representatives, there was at least partial support for this Dawson member's case in the community. After assessing the likelihood of successfully defending the new regulation and before the case went to court, the federal Department of Justice recommended a "voluntary compliance" approach rather than enforcement of a formal hunting regulation. In September 2007 the Yukon minister of environment announced to the PCMB that the one-week closure and the no-hunting corridor would not be enforced until further consultation had established consensus between all affected parties.

The political context changed dramatically after land claims agreements in the Yukon and the Northwest Territories were signed in the 1980s and 1990s, which added to an intercommunity political dynamic of First Nations autonomy. Whereas communities of the Northwest Territories fell under the Inuvialuit Final Agreement in 1984 and under the land claims agreements of the Gwich'in in 1992 (Government of Canada & Gwich'in Tribal Council, 1992), the Yukon First Nations, represented by the Council for Yukon Indians, negotiated and signed the Yukon Umbrella Final Agreement in 1993 (Government of Canada, Government of the Yukon Territory, & Yukon First Nations, 1993). Unlike the Northwest Territories land claims agreements, the Yukon agreement allowed self-government and regulatory organizations under separate First Nation agreements for each community. Eleven First Nations of the Yukon signed individual land claims and self-government agreements from 1995 to 2006. Each First Nation of the Yukon engaged in regional policy making while at the same time asserting its political autonomy.

The "let the leaders pass" initiative was in sharp contrast with the approach used by regional governments before the PCMB was created. It was less top-down and less restrictive for hunters compared with the loss of Indigenous hunting rights envisioned by some government agency officials of the pre-PCMB period. However, enforcement of the TEK-based rule was superseded by an assertion of an individual's right to hunt and was discontinued.

During its September 2007 meeting, the PCMB discussed the legal challenges that had undermined the formal "let the leaders pass" regulation in the Yukon Territory and endorsed voluntary compliance. This shift suggested that the emergence of greater First Nation political autonomy would necessitate a new role for co-management organizations in the future, the focus being more on facilitation than on making recommendations about specific regulations to government ministers. In subsequent efforts to develop a harvest management strategy, the PCMB strived to put a greater emphasis on achieving full consensus of parties (PCMB, 2010).

Barriers to Using TEK-based Hunting Regulations in Co-management

The efforts of the PCMB to experiment and learn from its experience with the Dempster Highway hunting issues demonstrate how co-management is a process of problem solving by trial and error and how, in integrating knowledge systems into co-management, the use of TEK can encounter significant barriers. Although co-management in some cases brings scientists, locals, and managers together for shared decision making (Armitage, Berkes, & Doubleday, 2007; Berkes & Berkes, 2009; Osherenko, 1988), combining knowledge systems is a challenging practice (Berkes, 2012; Clark, Clark, & Dowsley, 2010; Kofinas, 2005; Nadasdy, 2003a; Natcher, Davis, & Hickey, 2005). We identify barriers that prevented the lasting implementation of a TEK-based regulation. This list of barriers is not exhaustive, and some of the elements overlap, but we argue that they are the most important and that their identification and study may help future co-management enterprises to become more effective.

The context-specific nature of TEK limits its Application in resource management regulations

TEK and traditional hunting practices are concerned not only with facts and observations about the environment but also with how people relate to each other and with how they should behave in society and their environment (Berkes, 2012; Collings, 1997; Natcher, Davis, & Hickey, 2005). Conventional resource management has in some cases failed to recognize the social, cultural, and worldview dimensions of TEK (Nadasdy, 1999, 2007; Natcher, Davis, & Hickey, 2005). The concept of caribou leaders likely arose based on hunters' long-term observations to ensure the availability of caribou for harvest and to increase chances of human survival. In addition, the notion of letting caribou leaders pass is part of the norms for respectful hunting behaviour derived from human-environment relations and is thus central to the worldview of Indigenous hunters of the Porcupine caribou herd (see phase one).

The different components of TEK, such as social norms and spiritual aspects, cannot be separated from what is perceived as "factual" information, as doing so strips TEK of its meaning (Cruikshank, 1998; Houde, 2007; Wilson, 1996). The literature shows that TEK is locally restrictive and in most cases does not apply outside of a given locality (Antweiler, 1998; Cruikshank, 1998). For example, Cruikshank's (1998) study of oral traditions illustrates how elders' stories and the knowledge they convey are highly individual. Moreover, TEK directly relates to practices of resource use and can vary widely between social groups, time, and place (Berkes, Colding, & Folke, 2000). As explained with regard to phase four, interviews with Dawson hunters indicated a perception that the formal regulation imposing a one-week closure poorly matched the traditional notion of caribou leaders because it was not practical in the new context of highway hunting. Investigating the traditional meaning of the highly context-dependent term "caribou leader" through interviews with elders and hunters, we found that caribou leaders were described in terms of individual hunting experiences, in the context of traditional community hunts, and on a local scale, with perceptions varying according to the time of caribou migration, location, ecological context, political context, and hunting experience, including the status of the person interviewed (Padilla, 2010). With such varying concepts of caribou leaders, it was challenging to create one locally endorsed and cohesive regulation across a vast region (Smith & Cooley, 2003). The following respondents perceived differences in TEK practice and education between Indigenous hunters in the Porcupine herd region:

> You can't paint every First Nation with the same brush. This First Nation here, it'll say it is very good. We've got no one around here that hunts like that. (Dawson respondent)

> People you interview, they will probably interview different from me because we learn from our people, sometime our dad, our grandfather. I go in the mountain with them and they tell story about animals and I do something, I shoot caribou. (Fort McPherson respondent)

For some community members, the regulation appeared inconsistent with the stories they had heard from their elders or with their personal experience hunting caribou. Moreover, the regulation was an oversimplification and generalization of knowledge that is complex and situational. Distinct worldviews, varying concepts of caribou leaders, context dependence on local elders, and individual experiences caused a regional "one size fits all" regulation like the one-week closure to be contested. The one-week closure addressed neither local and individual variations nor social norms related to traditional knowledge of the "let the leaders pass" concept. However, the context-specific aspects of TEK alone did not preclude its use for management; rather, social context, political issues, and an institutional misfit prevented its use.

*Changes in systems of traditional authority, hunting technology,
and the social organization of harvesting caribou affect
the effectiveness of TEK approaches in
a contemporary social setting*

Hunting practice has become more individualistic (Barnhardt & Kawagley, 2005). Feldman (1997) suggests that traditional community-oriented authority systems have eroded. Elders interviewed echoed this assertion in terms of caribou hunting because there is no longer "one boss" organizing community hunts (phase one), and dissatisfaction among elders is ongoing because of unethical hunting within communities (phase four). Elders explained how the "let the leaders pass" rule was applied and respected under the system of traditional hunting authority. Similar accounts tell how the collective caribou hunt was well coordinated, with individuals having specific roles (Bali & Kofinas, 2008). Today, hunters of the Porcupine caribou herd are not always subject to the judgment of their peers or the oversight of elders during hunts that are carried out alone. The local initiative of the Fort McPherson traditional leader to address problems with highway hunting proved to be limited (phase three). Hunters also have easier and faster access to caribou with trucks and all-terrain vehicles along the highway. The shorter, individual hunts, less oversight by traditional leaders, and easier access to caribou make up the modern hunting context.

TEK regardng the "let the leaders pass" practice was incorporated into management outside of its traditional enforcement context of community-led hunts. TEK inspired the "let the leaders pass" policy in the context of adapting to modern hunting. The policy was not enforced in the context of social hunts, nor was it applied in the context of ensuring people's immediate survival. Letting the leaders pass was traditionally practised to ensure harvest by letting caribou settle in an area. The one-week closure to let the leaders pass was concerned with conservation of the Porcupine caribou herd and with an effort by elders to promote ethical hunting. Maximizing harvest was not central to the policy, although letting caribou settle in the area may lead to a higher harvest.

We argue that the modern hunting context created a barrier to applying the traditional notion of letting the leaders pass. Unlike the traditional practice of letting leaders pass, the formal, agency-enforced regulation was not highly flexible because the one-week closure did not call for hunters to judge during every hunt which caribou were the leaders of the herd. Indigenous hunters rely on TEK to guide their own hunts and to decide during the hunt whether leaders have passed. At the same time, the effectiveness of elders' informal enforcement of the "let the leaders pass" practice on a broad scale is limited because of the modern individual hunting context. In addition, the government's agency is limited by the guarantee of First Nation rights through land claims. Thus a formal regulation, even if based on TEK, is insufficient to ensure broad compliance with management plans or ethical hunting practices.

The Yukon government, following the recommendation of the PCMB, has modified Dempster Highway hunting regulations several times, fine-tuning them and trying to identify robust policies that will accommodate all user groups. Agrawal (1995) argues that, similar to science, TEK is ever-changing and thus not conserved indefinitely. TEK has two faces, one concerned with ancestral practice and the other concerned with adaptation to the present and future (Berkes, Colding, & Folke, 2000). TEK is therefore not bound to the past but is rooted in ancestral practices while incorporating aspects of modern societies (Stevenson, 1996). Local elders' revival of the "let the leaders pass" practice and the efforts to apply the concept through the PCMB management process show that TEK is adapting to modern hunting. Although the hunting closure to let the leaders pass did not last as an enforceable regulation, the renewed awareness of applying the traditional concept to highway hunting can be considered one step in the learning process.

Indigenous efforts toward self-government and political autonomy limit regional co-management consensus in a heterogeneous cultural landscape

In the regulatory regime during this research, First Nation members had an Indigenous right to hunt Porcupine caribou, unless there was a conservation issue (Government of the Yukon Territory, 2002). Collings (1997) argues that if political autonomy overrides concerns about resource availability, there will be a lack of consensus among users. A single Dawson First Nation hunter legally challenged the no-hunting closure to let the leaders pass, claiming inconsistency with his individual Aboriginal right based on the Tr'ondëk Hwëch'in (Dawson) First Nation's land claims agreement (PCMB, 2007). This agreement, signed in 1998 after other agreements had been signed and implemented, may help to explain this First Nation's high interest in asserting political autonomy and countering the formal closure intended to let caribou leaders pass.

In the past, the PCMB had enabled recognition of Indigenous people's hunting and wildlife management rights through its communications and recommendations. However, in the new context of settled land claims agreements and First Nations seeking to exercise their powers, there existed an institutional mismatch between the PCMB and the emerging Indigenous self-governments. In the case of the "let the leaders pass" discontinuation, formal recognition of a First Nation through its land claims agreement enabled an individual to defend his Indigenous hunting rights where a conflict arose with a PCMB recommendation. The Dawson First Nation perceived Dempster Highway hunting regulations imposed by the Yukon Territory government as interfering with its recently acquired political and cultural autonomy through self-government.

Adding to the barrier associated with greater political autonomy is the representation problem in northern co-management, which is well documented in other studies (Kofinas, 1998; Kruse, Klein, Braund, Moorehead, & Simeone, 1998). The representation problem results from board members who typically do not view themselves as speaking for their communities or First Nations, which in turn do not feel represented by the co-management board. This creates a communication disconnect in conveying the political position of First Nations during deliberations at board meetings.

Intercommunity conflict in the context of Dempster Highway hunting, in addition to an intergenerational divide between elders and younger hunters within communities, contributed to noncompliance and members challenging the regulation. Elders in interviews expressed concern about the younger generations' poor hunting ethics. Fort McPherson hunters historically had a bad reputation due to their highway hunting practices, whereas Old Crow hunters were the most highly regarded for having retained their traditional practice. An effort resulted on the part of elders across communities to promote traditional hunting practices, with a focus on the Dempster Highway. Although younger hunters reported hunting according to elders' teachings, they were concerned with defending their individual Indigenous rights in the face of government regulations. This historical context helps to explain their perspective. After the construction of the Dempster Highway, the focus of managers was on limiting Indigenous hunting along the highway. Although Indigenous hunters became part of the management process with the establishment of the PCMB, some hunters were still wary of formal management limiting their rights. When elders worked with the PCMB to create the hunting closure intended to let leaders pass, it was perceived by some younger hunters as an additional limitation on their Indigenous rights since it also restricted their ability to make the call when it came to letting caribou leaders pass. Elders' efforts to improve the younger generations' hunting practices were challenged because of the desire of younger hunters to affirm their Indigenous rights and what they perceived to be their own good hunting practices.

Through land claims agreements, First Nation members are empowered to become managers, yet formal regulations recommended through co-management still distinguish users from managers. The power of individuals to challenge decisions made through the co-management process decreases the potential for durable formal regulations and compliance. Consequently, users may perceive a co-management board as less valuable to advancing individual rights (Usher, 1993). Although TEK-based information may vary depending on the purpose and interests it serves, it is not neutral and can be used for political objectives (Agrawal, 1995; Wavey, 1993). PCMB management based solely on TEK exacerbated existing community-to-community conflict where there was a struggle between political autonomy and restrictive formal regulations that the board viewed as being for the common good. One Dawson hunter forecast the issue of compliance and intercommunity conflict: "Don't attack a person that doesn't go under YTG [Yukon Territory government] law because, like I said, it's going to lead into Indian wars, which is beginning now. It's going to come." The use of formal regulations based only on TEK exposed the local and regional heterogeneity of rules on caribou hunting. Ostrom (1990) reports that the absence of locally derived rules and enforcement, such as a lack of explicit recognition of land claims agreements and an unwillingness to recognize the legitimacy of diverging interests in a heterogeneous region, is an institutional weakness that can lead to institutional failure and ecological and social degradation. The need for explicit recognition of locally specific Indigenous rights and government through land claims agreements created an additional barrier because Indigenous people's traditional hunting practices are not homogenous. As a result, hunters resisted and contested regulations and enforcement that limited Indigenous rights in the context of efforts toward political autonomy.

The mismatch of agency enforcement of hunting regulations and TEK-based education is problematic

Elders started all interviews by talking about respectful hunting according to traditional values and their own elders' teachings. These respondents blamed the lack of good hunting practice on the absence of traditional education, not on a shortage of regulation enforcement:

> Well, what I think is sad, nobody teaching anybody. That's why they go up there and just shoot. Just one but you can't handle. There is no teaching going on. Well, all these things is ours, not for you, not for the government. It's our people doing that, it's up to us. See, you can't fix it ... But you have to teach them [how to] skin caribou and stuff like that, and we're not doing that. It's our fault, can't blame the younger people with it, it's our fault. We're not doing it. Elders supposed to tell chief what to do, not him. (Fort McPherson respondent)

Whereas some argue that traditional knowledge is not compromised but still fully practised with the younger generations (e.g., Sherry & Myers, 2002), others argue that older elders are the only members of the society left with the "original" traditional knowledge (Davis & Wagner, 2003). As Barnhardt and Kawagley (2005) note, researchers prefer capturing the knowledge of elders because the knowledge of the younger generations is regarded as eroded. The PCMB likewise relied on elders' knowledge of caribou leaders to derive a widely applicable highway hunting regulation. Padilla (2010) reports that elders' descriptions of caribou leaders are richer than younger hunters' versions, although the ecological knowledge of elders varies locally. Younger Dawson First Nation hunters in our interviews perceived the one-week closure as mismatched with their own elders' teachings and thus did not recognize it as legitimate.

The focus on enforcing this regulation regionally may have overshadowed the educational potential of the "let the leaders pass" policy. Some argue that conservation laws erode the social power of elders, placing them in an inferior social position (Phuthengo & Chanda, 2004). Some emphasize the use of traditional knowledge as an educational tool to promote sustainable harvest, not to produce prohibiting laws (Moller, Berkes, Lyver, and Kislalioglu 2004). Likewise, some authors caution against regulation based on TEK, focusing instead on gathering information and educating (Fernandez-Gimenez, Hays, Huntington, Andrew, & Goodwin, 2008; Wavey, 1993). During the PCMB's Dempster Highway Workshop, elders urged the revival of traditional hunting practices. Elders interviewed for this study warned that the younger generations were no longer following traditional knowledge and practices properly. Elders thus pushed for the "let the leaders pass" practice to ensure continued availability of caribou but also to bring youth back to traditional ways by addressing both conservation and social goals. This rationale and this strategy emphasized the social role of elders as educators.

Elders are considered the best educators for hunting because they have the most knowledge acquired throughout their lifetime (Hart, 1995). Fernandez-Gimenez et al. (2008) suggest that elders favour education over formalized hunting rules. Educating youth about TEK or traditional hunting practices, however, does not necessarily match modern practices but emphasizes values that were continually important to people's survival in the past (Barnhardt & Kawagley, 2005). The effort to bring the "let the leaders pass" rule to the forefront of management discussions was an educational one. Elders hoped to improve hunting practice along the highway. They addressed concerns about keeping caribou in the area rather than deflecting the herd from its wintering grounds. In our interviews, 83 percent of the respondents agreed that caribou leaders existed and described them according to hunting practice (Padilla, 2010). However, the formal hunting closure was not well received by all Indigenous hunters. A hunter described the confusing aspect of managing resource use through co-management boards and government-initiated regulations rather than elders' education:

> Natives use common sense, common knowledge. If they think and speak from their heart, they'll never have a problem. If you start speaking from your head in these boards and committees out there, governments, and pretty soon you're all mixed up and you don't even know what the decision is anymore. But if they think about it, go back to their elders, and these kids are smart now, in both worlds, I think they can take anybody on. (Dawson respondent)

Although the one-week closure was primarily based on a broad definition of caribou leaders derived from elders' knowledge, it did not actively engage elders in implementing the "let the leaders pass" rule locally. Thus elders blamed poor hunting practices along the Dempster Highway on the lack of education, whereas management focused on enforcement of the "let the leaders pass" regulation.

* * *

This case of a regulated "let the leaders pass" policy illustrates how the use of TEK, like science-based management of an earlier era, interfered with subsistence hunting and created a level of intercommunity conflict, a lack of compliance, and no long-term solution to a specific wildlife management concern. The one-week closure addressed neither local and individual variations nor aspects of social norms regarding the "let the leaders pass" concept. Some hunters perceived Dempster Highway hunting regulations imposed by the Yukon Territory government as interfering with recently acquired political autonomy through self-government.

Based on the findings of this case study, we make the following suggestions for future co-management. First, co-managers should be attentive to the differences between informal customary hunting traditions and formal rules that carry the force of law. Co-managers also need to ensure that recommendations generated by a co-management board have the full support of the stakeholders who share authority in resource governance. Where there is limited consensus, more consultation may be needed before regulatory actions are taken. This suggestion also comes with the acknowledgment that protracted conflict and inaction over a long period may in some cases threaten the ecological sustainability of a resource. Moreover, co-managers should find ways of supporting the role of elders in education and in the oversight of hunting activities. Finally, all parties should hold open discussions about the similarities and differences between their perspectives on traditional hunting practices (e.g., what a caribou leader is) to avoid or at least recognize conceptual ambiguity.

Hunters and wildlife managers perceived two benefits of the "let the leaders pass" policy: (1) ensuring the availability of caribou to hunters and (2) minimizing disturbance to herd migration, while promoting maximum use of the winter range by caribou. Ironically, an unintended consequence of letting leaders pass may be an increase in total harvest levels. Although deflecting caribou could lower harvest, other issues, such as the population effects of selectively eliminating leaders, could also be problematic (Padilla, 2010).

Although at some levels the implementation of the TEK-based regulation through co-management can be viewed as a failure, it can also be understood as one step in an ongoing social learning process of experimentation and adjustment based on experience. What appears to be a failure to find a workable solution to Dempster Highway hunting issues throughout the PCMB's history is at least in part the result of social-ecological change and an effort to adapt through the co-management process. Indeed, wildlife management approaches need to be flexible, allowing for changes in strategies to reflect the ever-changing and adaptive qualities of TEK and the social conditions in which they are embedded. The PCMB has encountered many challenges to its proposed strategies over the years. Nevertheless, the organization has succeeded in experimenting with new recommendations, learning from its experience, and moving on to implement new strategies, including rethinking its core functions as a co-management board. The barriers to letting the leaders pass should be seen as a door to adaptation and innovation for more effective future management of hunting on the Dempster Highway.

Acknowledgments

A version of this chapter appeared in *Ecology and Society* 19, 2 (2014).

We extend our thanks to members of the Porcupine Caribou Management Board who informed our study of Dempster Highway caribou-hunting management; to those interviewed, including Neil Colin, Art Christiansen, Woodie Elias, Amos Francis, Donald Frost, Percy Henry, Ronald Johnson, Roberta Joseph, Angie Joseph-Rear, Peter Josie, Danny Kassi, David Harvey Kassi, Peter Kay, Peggy Kormendy, Irwin Linklater, George S. Moses, Peter Nagano, Michael J. Pascal, Wanda Pascal, Joel Peter, Abraham Peterson, Robert Rear, Abraham Stuart, Randall Tetlichi, William Teya, Peter Tizha, Ernest Vittrekwa, and others who did not wish to be mentioned; to May and James André, Cheryl Charlie, James McDonald, and Ryan Peterson, each of whom assisted in interviews; to Joe Tetlichi and family for hosting Elisabeth Padilla in Old Crow; and to Archana Bali, Gregory Finstad, Brad Griffith, and Naomi O'Neil for comments on various editions of the chapter. This project was supported by the Porcupine Caribou Management Board, the Vuntut Gwitchin First Nation, the Tetlit Gwich'in First Nation, the Tr'ondëk Hwëch'in First Nation, the Government of the Yukon Territory, the Resilience and Adaptation Program at the University of Alaska Fairbanks, the Institute for Global

Awareness, and the US National Science Foundation, which funded the project Heterogeneity and Resilience of Human-Rangifer Systems: A Circumpolar Social-Ecological Synthesis. We also thank Nancy Tarnai for reading the chapter. The ideas and opinions expressed herein are those of the authors, not the funders.

References

Agrawal, A. (1995). Dismantling the divide between indigenous and scientific knowledge. *Development and Change, 26*(3), 413–39.

Alaska Department of Fish and Game. (2005). *Western Arctic Caribou Trails 7.* Juneau: Division of Wildlife Conservation, Alaska Department of Fish and Game.

Alaska Highway Pipeline Panel. (1979). *Initial Impact Assessment: Dempster Corridor.* Winnipeg: Alaska Highway Pipeline Panel.

Antweiler, C. (1998). Local knowledge and local knowing: An anthropological analysis of contested "cultural products" in the context of development. *Anthropos, 93*(4), 469–94.

Armitage, D., Berkes, F., & Doubleday, N. (Eds.). (2007). *Adaptive Co-management: Collaboration, Learning, and Multi-level Governance.* Vancouver: UBC Press.

Bailey, D.W., Dumont, B., & WallisDeVries, M.F. (1998). Utilization of heterogenous grasslands by domestic herbivores: Theory to management. *Annales de Zootechnie, 47*(5–6), 321–33. https://doi.org/10.1051/animres:19980501

Bali, A., & Kofinas, G. (2008). *Voices of the Caribou People.* Akureyri, Iceland: Conservation of Arctic Flora and Fauna. http://voicesproject.caff.is.

Barnhardt, R., & Kawagley, A. (2005). Indigenous knowledge systems and Alaska Native ways of knowing. *Anthropology & Education Quarterly, 36*(1), 8–24.

Baskin, L.M. (1989). Herding. In R.J. Hudson, K.R. Drew, & L.M. Baskin (Eds.), *Wildlife Production Systems: Economic Utilization of Wild Ungulates* (pp. 187–96). Cambridge, UK: Cambridge University Press.

Baskin, L.M. & Hjalten, A. (2001). Fright and flight behavior of reindeer. *Alces: A Journal Devoted to the Biology and Management of Moose, 37*(2), 435–45.

Benn, B. (2001). *Fall Movements of the Porcupine Caribou Herd Near the Dempster Highway, August 2000.* Inuvik: Gwich'in Renewable Resources Board.

Berger, T. (1977). *Northern Frontier, Northern Homeland: The Report of the Mackenzie Valley Pipeline Inquiry.* Ottawa: Minister of Supply and Services Canada.

Berkes, F. (1993). Traditional ecological knowledge in perspective. In J.T. Inglis (Ed.), *Traditional Ecological Knowledge: Concepts and Cases* (pp. 1–9). Ottawa: Canadian Museum of Nature.

Berkes, F. (2012). *Sacred Ecology: Traditional Ecological Knowledge and Resource Management* (3rd ed.). Philadelphia: Taylor and Francis.

Berkes, F., & Berkes, M.K. (2009). Ecological complexity, fuzzy logic, and holism in indigenous knowledge. *Futures, 41*(1), 6–12. https://doi.org/10.1016/j.futures.2008.07.003

Berkes, F., Colding, J., & Folke, C. (2000). Rediscovery of Traditional Ecological Knowledge as adaptive management. *Ecological Applications, 10*(5), 1251–67.

Berkes, F., & Turner, N.J. (2006). Knowledge, learning and the evolution of conservation practice for social-ecological system resilience. *Human Ecology, 34*(4), 479–94. https://doi.org/10.1007/s10745-006-9008-2

Carlsson, L., & Berkes, F. (2005). Co-management: Concepts and methodological implications. *Journal of Environmental Management, 75*(1), 65–76. https://doi.org/10.1016/j.jenvman.2004.11.008

Caulfield, R. (1997). *Greenlanders, Whales and Whaling: Sustainability and Self-Determination in the Arctic.* Lebanon, NH: University Press of New England.

Caulfield, R., Haley, S., Håkon Hoel, A., Hovelsrud-Broda, G., Jessen, A., Johnson, C., & Klokov, K. (2004). Resource governance. In N. Einarsson, J.N. Larsen, A. Nilsson, & O.R. Young (Eds.), *Arctic Human Development Report* (pp. 121–138). Akureyri, Iceland: Stefansson Arctic Institute.

Chapin, F.S., Kofinas, G., & Folke, C. (Eds.). (2009). *Principles of Ecosystem Stewardship: Resilience-Based Natural Resource Management in a Changing World.* New York: Springer.

Clark, D.A., Clark, S.G., & Dowsley, M. (2010). It's not just about bears: A problem-solving workshop on Aboriginal peoples, polar bears, and human dignity. *Arctic, 63*(1), 124–127.

Collings, P. (1997). The cultural context of wildlife management in the Canadian North. In E.A. Smith & J. McCarter (Eds.), *A Contested Arctic: Indigenous People, Industrial States, and the Circumpolar Environment* (pp. 13–40). Seattle: University of Washington Press.

Conradt, L., & Roper, T.J. (2005). Consensus decision making in animals. *Trends in Ecology & Evolution, 20*(8), 449–456.

Couzin, I.D., Krause, J., Franks, N.R., & Levin, S.A. (2005). Effective leadership and decision-making in animal groups on the move. *Nature, 433*(7025), 513–516. https://doi.org/10.1038/nature03236

Cruikshank, J. (1998). *The Social Life of Stories: Narrative and Knowledge in the Yukon Territory.* Lincoln: University of Nebraska Press.

Dahle, B.E., Reimers, E., & Colman, J.E. (2008). Reindeer (*Rangifer tarandus*) avoidance of a highway as revealed by lichen measurements. *European Journal of Wildlife Research, 54*(1), 27–35. https://doi.org/10.1007/s10344-007-0103-5

Davis, A., & Wagner, J. (2003). *Who* knows? On the importance of identifying "experts" when researching local ecological knowledge. *Human Ecology, 31*(3), 463–489. https://doi.org/10.1023/A:1025075923297

Dempster Highway Working Group. (1978). *A Dempster Highway Interim Management Proposal.* Whitehorse: Northern Roads and Airstrips Division, Department of Indian and Northern Affairs Canada.

Dumont, B., Boissy, A., Achard, A.M., Sibbald, A.M., & Erhard, H.W. (2005). Consistency of animal order in spontaneous group movements allows the measurement of leadership in a group of grazing heifers. *Applied Animal Behaviour Science, 95*(1–2), 55–66. https://doi.org/10.1016/j.applanim.2005.04.005

Feldman, S.P. (1997). The revolt against cultural authority: Power/knowledge as an assumption in organization theory. *Human Relations, 50*(8), 937–955. https://doi.org/10.1177/001872679705000804

Ferguson, M.A.D., & Messier, F. (1997). Collection and analysis of Traditional Ecological Knowledge about a population of Arctic tundra caribou. *Arctic, 50*(1), 17–28. https://doi.org/10.14430/arctic1087

Fernandez-Gimenez, M.E., Hays, J.U., Huntington, H.P., Andrew, R., & Goodwin, W. (2008). Ambivalence toward formalizing customary resource management norms among Alaska Native beluga whale hunters and Tohono O'odham livestock owners. *Human Organization, 67*(2), 137–150.

Folke, C. (2004). Traditional knowledge in social-ecological systems. *Ecology and Society, 9*(3), 7. Retrieved from http://www.ecologyandsociety.org/vol9/iss3/art7

Frank, J., & Frank, S. (1995). *Neerihiinjik: We Traveled from Place to Place*. Fairbanks: Alaska Native Language Center.

Fraser, M. (1976). Highway could deplete herd: Drastic steps sought for caribou. *Whitehorse Star*, May 21.

Geertz, C. (1973). Thick description: Toward an interpretive theory of culture. In C. Geertz (Ed.), *The Interpretation of Cultures: Selected Essays* (pp. 3–30). New York: Basic Books.

Glaser, B.G., & Strauss, A.L. (1967). *The Discovery of Grounded Theory: Strategies for Qualitative Research*. New York: Aldine.

Government of Canada, Government of the Yukon Territory, Government of the Northwest Territories, Council for Yukon Indians, Inuvialuit Game Council, Dene Nation, & Métis Association of the Northwest Territories. (1985). *Porcupine Caribou Management Agreement*. Ottawa: Department of Indian and Northern Affairs.

Government of Canada, Government of the Yukon Territory, & Yukon First Nations. (1993). *Umbrella Final Agreement between the Government of Canada, the Council for Yukon Indians and the Government of the Yukon*. Ottawa: Indigenous and Northern Affairs Canada. http://www.aadnc-aandc.gc.ca/eng/1297278586814/1297278924701.

Government of Canada & Gwich'in Tribal Council. (1992). *Gwich'in Comprehensive Land Claims Agreement*. Ottawa: Indigenous and Northern Affairs Canada.

Government of the Yukon Territory. (2002). *Wildlife Act: Revised Statutes of the Yukon*. Whitehorse: Government of the Yukon Territory.

Greer, S.C., & LeBlanc, R.J. (1992). *Background Heritage Studies: Proposed Vuntut National Park*. Gatineau, QC: Northern Parks Establishment Office, Canadian Parks Service.

Gubser, N.J. (1961). *Comparative study of the intellectual culture of the Nunamiut Eskimos at Anakuyuk Pass, Alaska* (PhD diss.). University of Alaska, Fairbanks.

Gunn, A., Arlooktoo, G., & Kaomayok, D. (1988). The contribution of the ecological knowledge of Inuit to wildlife management in the Northwest Territories. In M.M.R. Freeman & L.N. Carbyn (Eds.), *Traditional Knowledge and Renewable Resource Management in Northern Regions* (pp. 22–30). Edmonton: Boreal Institute for Northern Studies.

Hart, E.J. (1995). *Getting Started in Oral Traditions Research: A Case Study in Applied Anthropology in the Northwest Territories*. Yellowknife: Government of the Northwest Territories.

Houck, O. (2003). Tales from a troubled marriage: Science and law in environmental policy. *Science, 302*(5652), 1926–1929. https://doi.org/10.1126/science.1093758

Houde, N. (2007). The six faces of traditional ecological knowledge: Challenges and opportunities for Canadian co-management arrangements. *Ecology and Society, 12*(2), 34. Retrieved from http://www.ecologyandsociety.org/vol12/iss2/art34

Huntington, H.P. (2000). Using traditional ecological knowledge in science: Methods and applications. *Ecological Applications, 10*(5), 1270–1274.

Huntington, H.P. (2005). "We dance around in a ring and suppose": Academic engagement with traditional knowledge. *Arctic Anthropology, 42*(1), 29–32. https://doi.org/10.1353/arc.2011.0101

Inglis, J.T. (Ed.). (1993). *Traditional Knowledge: Concepts and Cases.* Ottawa: International Program on Traditional Ecological Knowledge, International Development Research Centre.

Ingold, T. (1986). Reindeer economies: And the origins of pastoralism. *Anthropology Today, 2*(4), 5–10. https://doi.org/10.2307/3032710

Kelsall, J.P. (1968). *The Migratory Barren-Ground Caribou of Canada.* Ottawa: Indigenous and Northern Affairs Canada and Canadian Wildlife Service.

Kendrick, A. (2003). Caribou co-management in northern Canada: Fostering multiple ways of knowing. In F. Berkes, J. Colding, & C. Folke (Eds.), *Navigating Social-Ecological Systems* (pp. 241–267). Cambridge, UK: Cambridge University Press.

Kofinas, G. (1998). *The costs of power sharing: Community involvement in Canadian Porcupine caribou co-management* (PhD diss.). University of British Columbia, Vancouver.

Kofinas, G. (2005). Caribou hunters and researchers at the co-management interface: Emergent dilemmas and the dynamics of legitimacy in power sharing. *Anthropologica, 47*(2), 179–196.

Kofinas, G. (2009). Adaptive co-management in social-ecological governance. In F.S. Chapin, G. Kofinas, & C. Folke (Eds.), *Principles of Ecosystem Stewardship: Resilience-Based Management in a Changing World* (pp. 77–101). New York: Springer.

Kofinas, G., Osherenko, G., Klein, D., & Forbes, B. (2000). Research planning in the face of change: The human role in reindeer/caribou systems. *Polar Research, 19*(1), 3–21.

Kruse, J., Klein, D., Braund, S., Moorehead, L., & Simeone, B. (1998). Co-management of natural resources: A comparison of two caribou management systems. *Human Organization, 57*(4), 447–458.

Light, S.S., Gunderson, L.H., & Holling, C.S. (1995). The Everglades: Evolution of management in a turbulent ecosystem. In L.H. Gunderson, C.S. Holling, & S.S. Light (Eds.), *Barriers and Bridges to the Renewal of Ecosystems and Institutions* (pp. 103–166). New York: Columbia University Press.

McCandless, R.G. (1985). *Yukon Wildlife: A Social History.* Edmonton: University of Alberta Press.

McClellan, C. (1987). *Part of the Land, Part of the Water: A History of the Yukon Indians.* Vancouver: Douglas and McIntyre.

McComb, K., Moss, C., Durant, S.M., Baker, L., & Sayialel, S. (2001). Matriarchs as repositories of social knowledge in African elephants. *Science, 292*(5516), 491–494. https://doi.org/10.1126/science.1057895

Miller, F.L., Jonkel, C.J., & Tessier, G.D. (1972). Group cohesion and leadership response by barren-ground caribou to manmade barriers. *Arctic, 25*(3), 193–202.

Moller, H., Berkes, F., Lyver, P.O., & Kislalioglu, M. (2004). Combining science and traditional ecological knowledge: Monitoring populations for co-management. *Ecology and Society, 9*(3), 2. Retrieved from http://www.ecologyandsociety.org/vol9/iss3/art2

Morrow, P., & Hensel, C. (1992). Hidden dissension: Minority-majority relationships and the use of contested terminology. *Arctic Anthropology, 29*(1), 38–53.

Nadasdy, P. (1999). The politics of TEK: Power and the "integration" of knowledge. *Arctic Anthropology, 36*(1–2), 1–18.

Nadasdy, P. (2003a). *Hunters and Bureaucrats: Power, Knowledge, and Aboriginal-State Relations in the Southwest Yukon.* Vancouver: UBC Press.

Nadasdy, P. (2003b). Reevaluating the co-management success story. *Arctic, 56*(4), 367–380.

Nadasdy, P. (2007). The gift in the animal: The ontology of hunting and human-animal sociality. *American Ethnologist, 34*(1), 25–43. https://doi.org/10.1525/ae.2007.34.1.25

Natcher, D., Davis, S., & Hickey, C. (2005). Co-management: Managing relationships, not resources. *Human Organization, 64*(3), 240–250.

Nellemann, C., Vistnes, I., Jordhøy, P., & Strand, O. (2001). Winter distribution of wild reindeer in relation to power lines, roads and resorts. *Biological Conservation, 101*(3), 351–60. https://doi.org/10.1016/S0006-3207(01)00082-9

Osherenko, G. (1988). *Sharing Power with Native Users: Co-management Regimes for Arctic Wildlife.* Ottawa: Canadian Arctic Resources Committee.

Ostrom, E. (1990). *Governing the Commons: The Evolution of Institutions for Collective Action.* New York: Cambridge University Press.

Padilla, E. (2010). *Caribou leadership: A study of traditional knowledge, animal behavior, and policy* (MSc thesis). University of Alaska, Fairbanks.

Page, R. (1986). *Northern Development: The Canadian Dilemma.* Toronto: McClelland and Stewart.

Paine, R. (1988). Reindeer and caribou *Rangifer tarandus* in the wild and under pastoralism. *Polar Record, 24*(148), 31–42. https://doi.org/10.1017/S0032247400022324

Paine, R. (1994). *Herds of the Tundra: A Portrait of Saami Reindeer Pastoralism.* Washington, DC: Smithsonian Institution Press.

Parlee, B., Manseau, M., & Łutsël K'é Dene First Nation. (2005). Using traditional knowledge to adapt to ecological change: Denésǫłıné monitoring of caribou movements. *Arctic, 58*(1), 26–37.

Perecman, E. & Curran, S. (Eds.). (2006). *A Handbook for Social Science Field Research: Essays and Bibliographic Sources on Research Design and Methods.* Thousand Oaks, CA: Sage.

Peter, A., & Urquhart, D. (1991). One caribou herd, two Native cultures, five political systems: Consensus management on the Porcupine caribou range. In Richard E. McCabe (Ed.), *Transactions of the Fifty-Sixth North American Wildlife and Natural Resources Conference* (pp. 321–25). Washington, DC: Wildlife Management Institute.

Phuthengo, T.C., & Chanda, R. (2004). Traditional ecological knowledge and community-based natural resource management: Lessons from a Botswana wildlife management area. *Applied Geography (Sevenoaks, England), 24*(1), 57–76. https://doi.org/10.1016/j.apgeog.2003.10.001

Pinkerton, E. (1989). Attaining better fisheries management through co-management: Prospects, problems, and propositions. In E. Pinkerton (Ed.), *Co-operative Management in Local Fisheries: New Directions for Improved Management and Community Development* (pp. 3–33). Vancouver: UBC Press.

Pinkerton, E. (1994). Conclusions: Where do we go from here? The future of traditional ecological knowledge and resource management in Native communities. In B. Sadler & P. Boothroyd (Eds.), *Traditional Ecological Knowledge and Modern Environmental Assessment* (pp. 51–60). Vancouver: Centre for Human Settlements, University of British Columbia.

Pinkerton, E. (1999). Factors in overcoming barriers to implementing co-management in British Columbia salmon fisheries. *Conservation Ecology*, 3(2), 2. Retrieved from http://www.consecol.org/vol3/iss2/art2

Pinkerton, E. (2009). Coastal marine systems: Conserving fish and sustaining community livelihoods with co-management. In F.S. Chapin, G. Kofinas, & C. Folke (Eds.), *Principles of Ecosystem Stewardship: Resilience-Based Natural Resource Management in a Changing World* (pp. 241–257). New York: Springer.

Porcupine Caribou Management Board (PCMB). (1995). *Porcupine Caribou Management Board Annual Report*. Whitehorse: PCMB.

Porcupine Caribou Management Board (PCMB). (2000). *Porcupine Caribou Management Board Annual Report*. Whitehorse: PCMB.

Porcupine Caribou Management Board (PCMB). (2006). *Porcupine Caribou Management Board Annual Report*. Whitehorse: PCMB.

Porcupine Caribou Management Board (PCMB). (2007). *Porcupine Caribou Management Board Annual Report*. Whitehorse: PCMB.

Porcupine Caribou Management Board (PCMB). (2010). *Harvest Plan for the Porcupine Caribou Herd in Canada*. Whitehorse: PCMB.

Rands, S.A., Cowlishaw, G., Pettifor, R., Rowcliffe, J.M., & Johnstone, R.A. (2003). Spontaneous emergence of leaders and followers in foraging pairs. *Nature*, 423(6938), 432–434. https://doi.org/10.1038/nature01630

Renaud, R. (1978). Promotion for Dempster area. *Northern Times* (Golspie, Scotland), September 8.

Richard, P.R., & Pike, D.G. (1993). Small whale co-management in the eastern Canadian Arctic: A case history and analysis. *Arctic*, 46(2), 138–143.

Sherry, E., & Myers, H. (2002). Traditional environmental knowledge in practice. *Society & Natural Resources*, 15(4), 345–358. http://www.tandfonline.com/doi/abs/10.1080/089419202753570828

Sherry, E., & Vuntut Gwitchin First Nation. (1999). *The Land Still Speaks: Gwitchin Words about Life in Dempster Country*. Old Crow, YT: Vuntut Gwitchin First Nation.

Singleton, S. (1998). *Constructing Cooperation: The Evolution of Institutions of Co-management*. Ann Arbor: University of Michigan Press.

Slobodin, R. (1962). *Band Organization of the Peel River Kutchin*. Ottawa: National Museum of Canada.

Slobodin, R. (1981). Kutchin. In J. Helm (Ed.), *Handbook of North American Indians, Volume 6, Subarctic* (pp. 514–32). Washington, DC: Smithsonian Institution Press.

Smith, B., & Cooley, D. (2003). *Through the Eyes of Hunters: How Hunters See Caribou Reacting to Hunters, Traffic, and Snowmachines near the Dempster Highway*. Whitehorse: Department of Environment, Government of the Yukon Territory.

Statistics Canada. (2006). *Canadian Census*. Ottawa: Government of Canada.

Stevenson, M. (1996). Indigenous knowledge in environmental assessment. *Arctic*, 49(3), 278–291.

Stewart, A.M., Keith, D., & Scottie, J. (2004). Caribou crossings and cultural meanings: Placing traditional knowledge and archaeology in context in an Inuit landscape. *Journal of Archaeological Method and Theory*, 11(2), 183–211. https://doi.org/10.1023/B:JARM.0000038066.09898.cd

Strauss, A.L. (1987). *Qualitative Analysis for Social Scientists*. New York: Cambridge University Press.

Taiepa, T., Lyver, P., Horsley, P., Davis, J., Bragg, M., & Moller, H. (1997). Co-management of New Zealand's conservation estate by Maori and Pakeha: A review. *Environmental Conservation, 24*(3), 236–250. https://doi.org/10.1017/S0376892997000325

Usher, P.J. (1986). *The Devolution of Wildlife Management and the Prospects for Wildlife Conservation in the Northwest Territories*. Ottawa: Canadian Arctic Resource Committee.

Usher, P.J. (1993). The Beverly-Kaminuriak caribou management board: An experience in co-management. In J.T. Inglis (Ed.), *Traditional Ecological Knowledge: Concepts and Cases* (pp. 111–120). Ottawa: Canadian Museum of Nature.

Warbelow, C., Roseneau, D., & Stern, P. (1975). The Kutchin caribou fences of northeastern Alaska and the northern Yukon. In D. Jakimchuck (Ed.), *Studies of Large Mammals along the Proposed Mackenzie Valley Gas Pipeline Route from Alaska to British Columbia*. Canadian Arctic Gas Ltd. and Alaskan Arctic Gas Study Company.

Wavey, R. (1993). International workshop on Indigenous knowledge and community-based resource management. In J.T. Inglis (Ed.), *Traditional Ecological Knowledge: Concepts and Cases* (pp. 11–16). Ottawa: Canadian Museum of Nature.

Wilson, S. (1996). *Gwich'in Native Elders: Not Just Knowledge, but a Way of Looking at the World*. Fairbanks: University of Alaska Press.

Wolfe, S.A., Griffith, B., & Gray Wolfe, C.A. (2000). Response of reindeer and caribou to human activities. *Polar Research, 19*(1), 63–73.

Yin, R.K. (2009). *Case Study Research: Design and Methods* (4th ed.). Thousand Oaks, CA: Sage.

14

Linking the Kitchen Table and Boardroom Table: Women in Caribou Management

Brenda Parlee, Kristine Wray, and Zoe Todd

(in memory of Elizabeth Colin)

The settlement of land claims in the western Arctic in the past three decades has led to the development of many new institutions of wildlife management in which Aboriginal people have taken leading roles. Among these are the co-management arrangements established for managing barren-ground caribou. Although these institutions have had their critics, they are largely considered among the best examples of co-management in the circumpolar Arctic in that they attempt to protect the rights and interests of Aboriginal peoples regarding the harvest of caribou, while at the same time ensuring the conservation of caribou populations (Berkes, 2009; Gunn, Russell, White, & Kofinas, 2009). However, many aspects of management exist outside of these institutions – informally organized at the local level by hunters and their families whose knowledge of caribou populations underlies many aspects of individual, household, and community well-being. Among the knowledge, practices, and beliefs that are little considered are those of Aboriginal women.

The archetype of "man the hunter" is deeply engrained in the archaeological and anthropological record of the North (Williamson et al., 2004). This bias has tended to become reproduced in the kinds of research carried out in the North as well as in policy contexts at regional, territorial, and national levels. Although some greater attention has been paid to gender biases since the 1970s, there is still relatively little in the academic literature that relates women and resources. Within this category, the theme of women and hunting is among the most complex. Although not well documented, women in many cultures play pivotal roles in hunting, ranging from performing spiritual rites to holding spears and other weapons (Geller & Stockett, 2007; Wadley, 2005).

The gender bias is not unexpected given that men were the sole leaders of anthropological tradition up until the mid-twentieth century (Parlee, Andre, & Kritsch, 2014). But the bias is also one that seems to have been easily and unapologetically replicated in other disciplines and research traditions, including that focused on traditional knowledge (Nadasdy, 2003). Whatever the reason, the end result has been a tendency to simplify knowledge systems of northern Aboriginal people as genderless. In this chapter, we attempt to take a more complex view, with an eye to discussing the knowledge, practices, and beliefs of Inuvialuit, Gwich'in, and Sahtú women. More specifically, what role do women play in the comings and goings of caribou? What kind of traditional knowledge is held by women? What role do women have in household and regional decisions about responses to caribou population change?

> I'm really happy to be in this gathering here with different communities just to learn from one another. I hardly go to meetings, and yesterday when we got into the hotel there, the manager said, "Oh you're here for the caribou meeting," and I said, "Yes, *we're* the hunters!" *[Laughter]* I think we brought it up last night, saying that this is the first time we have had a women and caribou meeting!

A workshop was held in Inuvik in September 2009 involving nine women from seven communities in the Sahtú, Gwich'in, and Inuvialuit regions: Fort Good Hope, Deline, Fort McPherson, Inuvik, Aklavik, Paulatuk, and Tuktoyaktuk. Graduate students working in four of the communities also participated in the meeting and guided discussion. The purpose of the meeting was to provide opportunities for women to share perspectives and experiences with one another related to caribou population change. The women were asked semi-structured questions related to their observations, experiences, and insights regarding how households were coping with the declining availability of caribou meat. The meeting was audio-recorded, and transcripts were prepared and shared with participants for verification. The content of the transcripts was subdivided into five themes, which are highlighted in this discussion.

Women and Changes in Availability of Caribou

Biologists over the past thirty to fifty years have offered numerous theories about the causes of declines in the barren-ground caribou population, ranging from habitat and predation models to broader theories of climatic change (Vors & Boyce, 2009). In many kinds of narratives from Gwich'in, Inuvialuit, and Sahtú elders, the caribou "come and go," which is how it has been for generations and generations. Although notions of cycles predominate, some oral histories also offer warnings about the future losses of caribou. During the workshop, elders offered cautionary reflections about the disappearance of the caribou drawn from previous predictions of elders and community prophets: "One day we're going to have nothing to work with. This is what our elders said. 'One day we're going to have no caribou, we're going to have no moose, we're going to have nothing.' And I really believe that. It's true." In both Inuvialuit and Dene oral trad-itions, the reciprocal relationship between people and caribou, including demonstrations of respect or lack of respect for caribou, is a key driver of population change and variation in distribution:

> A long time ago there was lots of caribou, at one time there was so much. It used to just come into the community. And people would just go out across the road and shoot a caribou. One day when the caribou came in, me and my husband said we would go and get a few caribou. We went down the side of the road and there was dead caribou lying everywhere with just their tongues cut out and their hind quarters taken. The rest they just left there. We went to the renewable resource officer and we told him and showed him. [Later] my dad said, do you know what they did with those caribou; they brought them to the dump and burnt them ... Since that day the caribou never have gone back to Fort Good Hope.

Women were integral to ensuring that harvested caribou were not wasted or treated in a disrespectful way:

> We were taught how to respect the caribou and the moose. It's very impor-tant that we use every part on a moose and caribou. Like when we used to move around like that, used to bring all of that, clean that, and guts and everything, and bring it home. And that's what we [go for]. Nothing is wasted. The skin, Mom used to cut the hair off. She tans the skin in the springtime. And then the caribou legs, she cleans them. She boiled the car-ibou hooves. The bone, she used to pound it up and make bone grease, too.

Previous anthropological investigations have highlighted the importance of women's power and influence over the hunt; some key rituals or rites of passage (e.g., young pubescent girls not touching hunting tools) directly implicate women in ensuring the caribou will come back to the people (Billson & Mancini, 2007; Brumbach & Jarvenpa, 1997; Cruikshank, 1979). Although some men hold similar medicine power, women are said to possess special powers of their own that they have to control. "The women accepted the responsibility to exercise this control and to behave in the appropriate manner; this was one of their indirect contributions to the hunters' work" (Heine, Andre, Kritsch, & Cardinal, 2001, p. 98). If women failed to do their part, the hunt might not be successful. Puberty was an important transition in women's development of their power. Only women had access to the knowledge. Therese Sawyer highlights how this knowledge was shared between women: "They taught me how to respect all those things that were associated with the way you lived in those years. But these things were brought to me by other women in the later years. By that time, you became a woman, so now these women were telling me I should be respecting the women" (quoted in ibid., p. 98).

Changes in the availability of caribou are thus social-ecological in nature; changes in the balance of respect between people and caribou, including demonstrations of a "lack of respect," are viewed as fundamental drivers of caribou population change. Women play a key role in maintaining this respectful relationship, which involves limiting waste by using all of the caribou, as well as rituals and rites of passage particular to women.

WOMEN AND TRADITIONAL KNOWLEDGE SHARING

Knowledge about caribou does not, however, exist in a fixed space and time. Women talk a great deal about teaching youth and future generations about how to respect caribou. Many women have shared stories about the past and their experiences learning from their parents and grandparents about what was important and how to survive:

> I work with the hides. Now I'm getting old and I'm struggling, but I still never give up. I really enjoy cutting meat, I'll tell you that. It makes me feel so good just to be sitting with a pile of meat. Like I know how to work with the meat. All week I've been doing that, I was so happy. My son got

meat he brought us, and me and my husband just sit together. He helps me. It just makes me feel so good just to do that. And then I would say to him, "Now we're okay for a while."

My children and grandchildren all get involved with what I do. I'm teaching them. I have got to have a lot of patience, too. I have always said that my mom never had patience with us, but Mom was a good teacher, too. My dad was such a patient man ... he would never get mad. Even when he had to repeat something to us ten times, he never got mad. But Mom only said things once.

The kinds of knowledge and skills held by Gwich'in women were described by workshop participants as distinct from those of men. For example, women are involved in many aspects of processing, storing, and preparing meat as well as hides. The preparation of hides is a particularly important responsibility and involves a great deal of work, particularly in summer months (see Table 14.1).

Preparing hides and sewing slippers, mitts, and mukluks are traditions that continue to be important in many communities. Women in the workshop raised concerns that passing on this knowledge becomes increasingly difficult within the context of a reduced caribou harvest. There is concern that youth today will no longer have the skills to work with caribou hides, a key to "not wasting caribou." As discussed above, this wasting equates to a lack of respect, which in turn is perceived as a fundamental driver of caribou population change.

Workshop participants also shared more encouraging stories about the way that certain programs are facilitating relationship building between elders and youth and leading to positive learning experiences. An elder emphasized the value of reinforcing and celebrating the successes of youth in hunting as well as in the practices of sharing:

Lots of kids I know are getting interested in [being on the land] now ... And when they go hunting, they bring the meat back [and] they give a piece to everybody, which is really good, too. They share with the community. And the elders see how the young people are cutting their meat up. And the peoples are all getting meat like that. I just wish they could talk on CBQM [local radio] and say thank you to these young people for doing such a good job, but I never hear anything like that. Just to encourage them. Our youth today have lots to learn. They need someone to teach them after we're gone.

TABLE 14.1   Example of phases of tanning caribou hides

A good time to work	Spring and summer
Soaking the skin.	Moose and caribou hides are soaked in a river or lake for about one week.
Scorching the hair side.	Before the water is wrung out, the hair side has to be scorched to take the little bits of hair off the hide. A special pan is heated and then placed upside down on the ground, and the hide is moved across it. When this step has been completed, the hide is black where it has been scorched.
Thinning out the hair side.	Next, the hair side around the neck area is thinned out because this is the thickest part of the hide.
Wringing out the water.	To drain out all of the blood that remains in the skin, the hide is wrung out and rinsed in clear water for the first time. For every rinsing, fresh water is required.
Twisting the wet hide.	To prepare the hide for twisting, small loops are cut out along the hide's outside edge. A solid stick is placed into the ground, and about three of the loops are slipped over it at the same time. Three loops on the opposite side of the hide are slipped over a sturdy piece of driftwood about 1 metre long. The pole is used for twisting the hide. After it is twisted, it is rinsed and then twisted again. This is continued until only clean water comes out when the hide is twisted.
Soaking the hide in warm brain water.	To soften the hide further, it is then soaked in brain water. After being soaked, it is twisted again, then soaked again, and then twisted again.
Scraping the hide on the hair side.	The hide is wrapped over a pole that lies between two tripods (about 1.5 metres off the ground). The hide is scraped on the hair side to remove the hair.
Scraping the hide on the inside.	The hide is scraped with a stone scraper until it is more or less dry, and then the dried flesh is removed using a metal scraper. To make the scraping easier, the horizontal pole is lowered so that it is about 1 metre above the ground.
Preparing the hide for smoking.	The hide is smoked to soften it for making moccasins, pants, and coats. Before it is smoked, holes are sewn up and edges are trimmed off. The hide is hung on a stick 1 to 1.25 metres off the ground, with a canvas sometimes attached to the bottom to act as a funnel for the smoke.
Smoking hides.	The smoking fire is made of rotten wood and has to be well tended so that it will not go out or catch on fire and scorch the hide. The smoking fire is kept burning for almost a whole day (8–10 hours). Once the inside of the hide is smoked, the hide is taken off and scraped to soften it.

Source: Adapted from Heine, Andre, Kritsch, & Cardinal (2001, p. 142).

HOUSEHOLD ECONOMY: DIET AND HEALTH

The majority of human and financial resources and discussion associated with managing caribou in the recent decade has focused on managing people – namely harvesters. There is little evidence that harvesting is of any concern (Gunn et al., 2011) in comparison to evidence showing the impacts of resource development on caribou habitat (Johnson et al., 2005).

There is no question, however, that harvesters and their families who depend on caribou meat as a dietary staple are the most impacted by declining caribou populations. With the declining availability of caribou, what kinds of food alternatives are available, and what choices are individuals and households making among those alternatives? The lack of attention to these fundamental questions of human health, particularly in light of growing concerns about diabetes and chronic illness among northern Aboriginal peoples, highlights the need for a more integrated social-ecological approach to northern resource management.

Within most of the households of the Inuvialuit, Gwich'in, and Sahtú communities, women have long played a primary role in ensuring the health and well-being of their families. Many families consider the practice of "caribou hunting" to be part of a healthy lifestyle:

> We raised our kids in the bush. One of my sons was born in September, and we went to the bush on September 21. Was just three weeks old, just newborn, and the nurse told me, "Mrs., you're crazy to take that newborn out in the bush." I said, "It's more healthy out there than it is in town." So anyway, we went to the bush, came back December 19, brought my little boy to the nurse. "Oh, I see you're healthy!" Yeah, really healthy, and then two weeks later, he was just so sick! *[Laughter]* Because the kids all had that cold in town, got really sick with that cold. When you're out in that bush, yeah, clean fresh air, and the kids never get sick.
>
> In order to have everything, we had to really work hard. But you know, we never thought it was working hard. We never got sick those days compared to today. Because there was lots of fresh air. *[Pause]* Lots of fresh meat.

"Hunting" goes beyond simply the act of killing the animal. Women, in particular, play important and fundamental roles in cooking, preserving meat, and preparing hides, as well as sewing hides into clothing and tools for use by the family. All of these responsibilities, as well as the harvest of the animal itself, are considered "hunting" in many communities, as discussed by Bodenhorn (1990) in her paper "I'm Not the Great Hunter, My Wife Is" (see also Parlee, Andre, & Kritsch, 2014).

However, the traditional lifestyle of caribou hunting has changed significantly over the past decades; in addition to carrying on aspects of caribou-harvesting traditions, women have become the primary wage earners in the majority of Inuvialuit, Gwich'in, and Sahtú communities. The aggregated average employment rate for women in the three regions is 48.9 percent, which is slightly higher than the average for men at 46.3 percent. If we exclude Inuvik and Norman Wells, towns where there are larger populations of non-Aboriginal peoples, the employment differential is much greater, with women having an employment rate of 44.1 percent versus 38.7 percent for men (NWT Bureau of Statistics, 2012). As both caribou hunters and wage earners, women are on the frontline of household adaptations to changes in the availability of food resources. With the declining availability of caribou, what kind of food choices are women and households making among traditional food and store-bought food as alternatives?

An understanding of the influence of caribou population change on food choices begins with an accounting of whether or not people perceive a decline in caribou numbers or consider conditions to be appropriate for continued harvesting (see Figure 14.1). Whereas some harvesters in some communities may see little change in the availability of caribou (a), others may be observing and experiencing a decline similar to what is reported

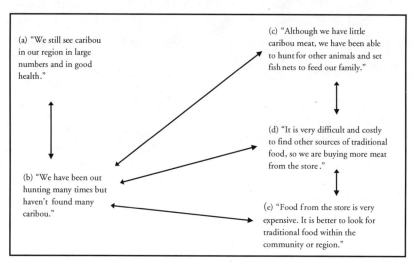

FIGURE 14.1   Food choices in the context of changing caribou populations

by government biologists (b). Those observing and experiencing a decline may respond in the short term by looking for caribou farther afield but over time are more likely to focus on the harvest of other species (c), to buy other store-bought meat instead, or to use social networks within the community and region to find other sources of traditional foods (d).

Because harvest management plans have typically lacked consideration of such socio-economic factors driving harvest behaviour, the impact of harvest management on communities is little understood, and we do not know the extent to which harvesters can or will comply with harvest regulations (Todd, 2010).

WOMEN IN CO-MANAGEMENT

Women play complex roles in managing caribou and their relationship to caribou. However, this knowledge has largely been absent from the formal processes of caribou co-management that are in place in the Inuvialuit, Gwich'in, and Sahtú regions. Few of the representatives on the six management boards charged with the management of the Porcupine, Bluenose West, Bluenose East, and Cape Bathurst herds are women (see Table 14.2); of the forty-four positions, only one (0.023 percent) was reported to be held by a woman.

TABLE 14.2 Aboriginal women as representatives in the co-management of the Porcupine, Bluenose East, Bluenose West, and Cape Bathurst caribou

Co-management board	Number of Aboriginal women as members	
Wildlife Management Advisory Council – North Slope	0/4	No Aboriginal women are currently members of the council (one alternate).
Porcupine Caribou Management Board	0/8	No Aboriginal women are currently members of the board.
Inuvialuit Game Council	0/8	No Inuvialuit women are currently members of the council.
Gwich'in Renewable Resources Board	0/8	No Gwich'in women are currently members of the board (one alternate).
Sahtú Renewable Resources Board	1/8	One Sahtúgótįnę woman is currently a member of the board.
Wek'èezhìi Renewable Resources Board	0/8	No Aboriginal women are currently members of the board.
Total	1/44 (0.023%)	

Although there are many factors known to influence population changes, including climate and habitat disturbance and degradation, these boards have almost singularly focused on the issue of caribou harvest. This issue has been a dominant concern of government for many decades; however, it may be an ecological red herring according to many scholars (see Chapter 3).

The more complicated issue in the context of caribou population decline is arguably the sustainability of families and communities dependent on caribou populations reported as in decline; social, economic, health, and culture are not themes regularly discussed at the boardroom table. The assumption behind such a narrow perspective is seemingly that individual choices and household decisions fall in the domain of personal responsibility, where the government should play no formal role. However, individuals and households are not in complete control of the environment that influences their food choices. In that context, critical consideration of the role of regional, territorial, and federal governments in addressing the social, economic, and health implications of ecosystem change may be needed.

What are the synergies between the kitchen table, where women play a leading role, and boardroom tables, which are overwhelmingly dominated by men? As highlighted by a Gwich'in woman and elder,

> I'm really glad I came and that women play a big part in the caribou, too, because, you know, they help the men, okay, but it's the women that do all the work when they bring the meat back. The women, at one time, women were not permitted to talk. Nothing! Nothing in meetings, you know? You go to meetings, you just listen. But today it is different. Aboriginal women are in different committees. Women weren't allowed to sit in meetings. Now they are there and they are talking. Which is good to see. So it'll be nice when we get home to share what we've learned, you know – that women do have voice for the caribou, too, not only men.

These issues were more than theoretical to the Inuvialuit, Gwich'in, and Sahtú women involved in the research program. There was a sense of urgency and responsibility that women need to be more active in dealing with the issues of caribou management:

> [Discussing the issues with women] kind of brought a different voice to issues of caribou. It's not a male voice; it's a female voice. And it's nice to know that a cross-section of ages here, that regardless of what age we're in,

we're all very connected to our culture and knowing that that's probably the key thing that's vital to whether the caribou are going to continue to be a part of our sustainability. And I think that the management of caribou is going to be something that we can't just brush aside. But our voice comes from a different place. We know that we've had concern, and we need to do more ... We are really part of the whole caribou hunt.

* * *

We ask readers to consider the voices of women in the debate on caribou population change and management in northern Canada. Like other theses on gender and resource management, we offer the hypothesis that women have unique roles and experiences with respect to caribou that in turn have generated unique knowledge about human-caribou relations. We recognize that attention to gender can bring insights about the differentiated impacts of resource management decisions on communities and households. It follows from this hypothesis and these assumptions that paying attention to the voices of Inuvialuit, Gwich'in, and Sahtú women is relevant to our understanding of not only "caribou" but also the institutions that have evolved to mediate human-caribou relations.

The voices of women shed light on the reality that human-caribou relations extend beyond the hunt itself. Rules about respectful behaviour in relation to caribou are important to Inuvialuit, Gwich'in, and Sahtú views of caribou population variability and change; there are particular rules of respect that implicate women in ensuring a successful hunt. There are also rules that guide the use and sharing of caribou meat within the household; being healthy is strongly related to the experience of hunting, or being on the land, as is the availability of fresh meat. The strongest themes and rules relate to teaching youth about respecting caribou and passing on traditional knowledge; ensuring opportunities for youth to be on the land is seen as fundamental to the continuation of cultural practices such as tanning caribou hides and the related work of women. In contrast, there are few rules or social norms that link women to formal institutions of management.

There are three issues or implications of this study that can be fruitfully expanded upon. The first issue relates to the attention paid to community as the main scale or unit of analysis; as noted by Agrawal and Gibson (1999) among others, there is a tendency to treat community, particularly Indigenous communities, as homogenous and simplified in values and knowledge rather than to recognize the complexity and diversity of ideas and voices that exists there. This rather simplified approach to community has thus obscured the ways that we understand both the impacts and responses of Indigenous peoples to caribou population change.

The lack of consideration of Inuvialuit, Gwich'in, and Sahtú women's voices also stems from biases in the historical ethnographic record. Early anthropologists in the western Arctic provided only limited details of women and their roles vis-à-vis resources, including caribou. Only generalized interpretations of women can be found. According to Jenness (1972, p. 403), the contribution of Gwich'in women to the overall livelihood of the community was more than significant, yet their status in the overall decision making was limited: "[They] received no gentle treatment; they performed nearly all the hard work in camp, transported all the family possessions, ate only after the men had eaten and had no voice in family or tribal affairs." This generalized view of women as disadvantaged in the structures and processes of decision making is not considered accurate by many contemporary scholars whose revisionist histories of northern peoples emphasize principles of equity and shared decision making and more complex kinds of female roles and responsibilities.

Overly romanticizing the female archetype sits at the other extreme. In this case, women are constructed as being closer to the earth and in tune with resources; such Rousseau-like constructions, although empowering on some level, are also akin to the idea of the "ecological Indian," which has been critiqued for its simplicity (Krech, 1999). Neither images of women as oppressed nor images of women as "mother earth" are useful in their extremes. Further research to address the gap in understanding is needed; the gap continues to exist with respect to our understanding of women's knowledge and experience of their environment and how in turn that knowledge and experience inform the kinds of rules that are in place and women's perceptions and use of resources.

The second issue relates to scale and the acknowledgment of the kinds of social, economic, and cultural factors that inform social-ecological relations and rulemaking. The majority of wildlife management decisions are oriented around or derived from conservation traditions that suffer harvesting as a necessarily political requirement of management rather than a facet of a healthy social-ecological system. This focus on conservation has led to the construction of harvest models and management plans that are concerned solely with ecological or biological parameters at the scale of the caribou range, and it tends to ignore social-ecological relations and the drivers of harvest behaviour. Many caribou management plans evolving today in the western Arctic strongly focus on rule making in this ecological context (as demarcated by the dark lines in Figure 14.2), ignoring

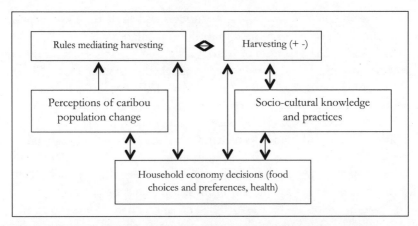

FIGURE 14.2   A model of the socio-economic and cultural drivers of harvesting

the more complex ways that rules are constructed and that rules influence human behaviour. Such orientation leaves little room for imagination or contemplation of harvesting as a socio-economic and cultural practice. It is precisely these social, economic, and cultural factors, however, that drive perceptions of caribou population change and in turn influence harvesting decisions. The model assumes a dynamic interrelation between harvesting, household economic decisions (including choices about the allocation of resources for food budgets and food preferences), and socio-cultural knowledge and practice.

Feedback between harvesting households' economy decisions, socio-cultural knowledge, and perceptions of caribou population change is also notably important. As caribou become less represented within the household diet, and choices are made to find alternative foods, the capacity to pass on to future generations traditional knowledge related to caribou harvesting is affected. This factor in turn influences how caribou population change is read and understood over time. One can expect that over time, traditional knowledge will become less represented in rulemaking, thus complicating harvesting.

Women play key roles in many aspects of this social-ecological system. Their role as decision makers in the economy of the household – which in a basic sense involves decisions about what goes on the kitchen table – is arguably among the most important to nutrition and health. In the absence of caribou meat, women are left with the more challenging task of identifying available and economically viable substitutions. The focus on women and the household does not necessarily mean that women are physically present as "homebodies." Employment statistics in the western Arctic and elsewhere in the North highlight women as being well represented in the wage economy. Although the territorial average shows similar employment rates for both men and women, in many communities, wage employment rates for women are higher than for men from the same community owing to a variety of factors, including education (NWT Bureau of Statistics, 2012). The number of women heading single-parent families is also much higher than the number of men. Even with the growing development of the resource sector, which includes mining and tends to employ more men than women (e.g., in construction work), the decisions about how to feed and care for the family, particularly children and elders, falls to women.

Conversely, there is significant underrepresentation of women in wildlife management, a problem that seems perpetuated by the lack of acknowledgment of women's roles and knowledge as relevant to our understanding not only of changing wildlife resources (i.e., caribou) but also of changing resource economies. The lack of presence of women in the boardroom has been described as irrelevant by some; the assertion that real power is held elsewhere or in the home is a usual response. This placation obscures both the obvious issue that what happens in the home should matter in policy and conversely the fact that greater attention might be paid to how regional, territorial, and global policy in turn influences what occurs in the home.

References

Agrawal, A., & Gibson, C.C. (1999). Enchantment and disenchantment: The role of community in natural resource conservation. *World Development, 27*(4), 629–649.

Berkes, F. (2009). Evolution of co-management: Role of knowledge generation, bridging organizations and social learning. *Journal of Environmental Management, 90*(5), 1692–702.

Billson, J.M., & Mancini, K. (2007). *Inuit Women: Their Powerful Spirit in a Century of Change.* Toronto: Rowman and Littlefield.

Bodenhorn, B. (1990). I'm not the great hunter, my wife is: Iñupiat and anthropological models of gender. *Études/Inuit/Studies, 14*(1–2), 55–74.

Brumbach, H.J., & Jarvenpa, R. (1997). Woman the hunter: Ethnoarchaeological lessons from Chipewyan life-cycle dynamics. In C. Claassen & R.A. Joyce (Eds.), *Women in Prehistory: North America and Mesoamerica* (pp. 17–32). Philadelphia: University of Pennsylvania Press.

Cruikshank, J. (1979). *Athapaskan Women: Lives and Legends.* Ottawa: National Museum of Man.

Geller, P.L., & Stockett, M.K. (2007). *Feminist Anthropology, Past, Present and Future.* Philadelphia: University of Pennsylvania Press.

Gunn, A., Johnson, C., Nishi, J., Daniel, C., Russell, D.E., Carlson, M., & Adamcze-wski, J. (2011). Understanding the cumulative effects of human activities on barren-ground caribou. In P.R. Krausman & L.K. Harris (Eds.), *Cumulative Effects in Wildlife Management: Impact Mitigation* (pp. 113–133). Boca Raton, FL: CRC Press.

Gunn, A., Russell, D., White, R., & Kofinas, G. (2009). Facing a future of change: Wild migratory caribou and reindeer. *Arctic, 62*(3), iii–vi.

Heine, M., Andre, A., Kritsch, I., & Cardinal, A. (2001). *Gwichya Gwich'in Googwandak: The History and Stories of the Gwichya Gwich'in as Told by the Elders of Tsiigehtchic, NT.* Tsiigehtchic and Yellowknife: Gwich'in Social and Cultural Institute.

Jenness, D. (1972). *The Indians of Canada.* Toronto: University of Toronto Press.

Johnson, C.J., Boyce, M.S., Case, R.L., Cluff, H.D., Gau, R.J., Gunn, A., & Mulders, R. (2005). Cumulative effects of human developments on Arctic wildlife. *Wildlife Monographs*, (160): 1–36.

Krech, S. (1999). *The Ecological Indian: Myth and History.* New York: W.W. Norton.

Nadasdy, P. (2003). *Hunters and Bureaucrats: Power, Knowledge, and Aboriginal-State Relations in the Southwest Yukon.* Vancouver: UBC Press.

NWT Bureau of Statistics (2012). *Community Data.* Yellowknife: Government of the Northwest Territories. http://www.statsnwt.ca/community-data/index.html.

Parlee, B., Andre, A., & Kritsch, I. (2014). Offerings of stewardship: Celebrating life and livelihood of Gwich'in women in the Northwest Territories. In C.R. Wilson & C. Fletcher (Eds.), *Native Peoples: The Canadian Experience* (4th ed.). Don Mills, ON: Oxford University Press.

Todd, Z. (2010). *Food security in Paulatuk, NT: Opportunities and challenges of a changing community economy* (MSc thesis). University of Alberta, Edmonton.

Vors, L.S., & Boyce, M.S. (2009). Global declines of caribou and reindeer. *Global Change Biology, 15*(11), 2626–2633. https://doi.org/10.1111/j.1365-2486.2009.01974.x

Wadley, S.S. (2005). *Essays on North Indian Folk Traditions.* New Delhi: Chronicle Books.

Williamson, K.J., Hoogensen, G., Lotherington, A.T., Hamilton, L.H., Savage, S., & Koukarenko, N. (2004). Gender issues. In N. Einarsson, J. Nymand Larsen, A. Nilsson, & O.R. Young (Eds.), *Arctic Human Development Report* (pp. 187–205). Akureyri, Iceland: Stefansson Arctic Institute.

Contributors

Leon Andrew is a Sahtú elder of Tulit'a, Northwest Territories, who is involved in many research and development programs in his community and region. He is also an elder adviser to the Sahtú Renewable Resources Board. He lives in Norman Wells, Northwest Territories.

Fikret Berkes is a professor of natural resource management at the University of Manitoba. He is an applied ecologist with a background in addressing issues of natural science and social science significance. Berkes has focused most of his research career on exploring the interrelationships between social and ecological issues or social-ecological systems. His expertise on the theory and co-management of common-pool resources is recognized globally.

Peter Boxall is a professor of resource economics in the Department of Resource Economics and Environmental Sociology at the University of Alberta. His research spans a range of themes related to the sustainability of natural resources in Canada and Australia.

Ken Caine is an assistant professor in the Department of Sociology at the University of Alberta. His work has mainly focused on the governance of natural resources in the Canadian North. A particular emphasis has been on the role of traditional knowledge in land-use planning and in the co-management of wildlife, fish, and water resources.

Robert Charlie is a Gwich'in harvester originally from Fort McPherson, Northwest Territories, and currently resides in Inuvik, Northwest Territories. He is a former chair of the Gwich'in Renewable Resources Board and the current director of the Education and Training Division of the Gwich'in Tribal Council.

Anne Marie Jackson is a Sahtú youth and lives in Tulit'a, Northwest Territories.

Norma Kassi is a former chief of the Vuntut Gwitchin First Nation and currently lives in Old Crow, Yukon.

Gary P. Kofinas is a professor of resource policy and management at the Institute of Arctic Biology and in the Department of Natural Resources, both at the University of Alaska Fairbanks. He has led numerous research initiatives in northern Canada, Alaska, and the circumpolar North on the resilience of human-caribou systems. He was a lead researcher of the Circum-Arctic Rangifer Monitoring and Assessment Network, funded by the International Polar Year program, and a principal investigator of the National Science Foundation project Heterogeneity and Resilience of Human-Rangifer Systems: A Circumpolar Social-Ecological Synthesis.

Tobi Jeans Maracle is a master's student at the University of Saskatchewan with significant experience in the Yukon Territory. Her research in Old Crow, Yukon, on food-sharing networks has contributed to the efforts of the Vuntut Gwitchin to deal with changes in the availability of barren-ground caribou and other traditional food resources.

Roger McMillan is a provincial researcher for the Government of British Columbia. He has a master of science degree in environmental sociology from the University of Alberta. His thesis work, carried out in collaboration with the community of Fort Good Hope, Northwest Territories, focused on the social networks of communities in the context of adaptation to changing caribou populations.

David Natcher is an associate professor and director of the Indigenous Land Institute at the University of Saskatchewan. His work focuses on a diversity of issues related to Aboriginal resource management and community development in northern Canada and Alaska.

Morris Neyelle is an elder and caribou hunter from the Sahtúgótįnę community of Délįnę. He is a well-known photographer and has played key roles in the leadership and governance of his community and in the management of barren-ground caribou. He currently lives in Délįnę, Northwest Territories.

Elisabeth Padilla completed her master of science degree in wildlife biology at the University of Alaska Fairbanks in 2010, with a focus on the caribou management practices of hunters in Dawson City and Old Crow, Yukon, and in Fort McPherson, Northwest Territories. She is currently a museum educator for the Museum of the North at the University of Alaska and lives in Fairbanks.

Brenda Parlee is an associate professor in the Faculty of Agricultural, Life, and Environmental Sciences at the University of Alberta. She has been working for twenty years in the Northwest Territories on many different issues of social and ecological change. Her key focus has been on the role of traditional knowledge in resource management processes, including monitoring.

Frank Pokiak is an Inuvialuit harvester of Tuktoyaktuk in the Inuvialuit Settlement Region, Northwest Territories. He is a former chair of the Inuvialuit Game Council. He lives in Tuktoyaktuk, Northwest Territories.

John Sandlos is an associate professor in the Department of History at Memorial University, Newfoundland, where he studies the history of the Canadian conservation movement and the environmental history of northern Canada. He is the author of *Hunters at the Margin: Wildlife Conservation in the Northwest Territories* (2006).

Glenna Tetlichi is a former band councillor of the Vuntut Gwitchin First Nation and currently lives in Old Crow, Yukon.

Zoe Todd is a professor at Carleton University. She has a master of science degree from the University of Alberta and a doctoral degree from the University of Aberdeen. Her research in Paulatuk, Northwest Territories, focused on how wage employment influences the interrelationships between food security and the changing employment patterns of Arctic communities. She currently lives in Ottawa.

Kristine Wray is a doctoral candidate in environmental sociology at the University of Alberta. She completed a master of science degree based on research with the Gwich'in Renewable Resources Board. She lives in Edmonton, Alberta.

Natalie Zimmer has a master of science degree in resource economics from the University of Alberta. Her thesis focuses on how perceptions of wildlife health in Alberta might affect hunting behaviours.

Index